THE GEOLOGY OF THE METAMORPHOSED BIWABIK IRON-FORMATION, EASTERN MESABI DISTRICT, MINNESOTA

UNIVERSITY OF MINNESOTA
MINNESOTA GEOLOGICAL SURVEY
PAUL K. SIMS, DIRECTOR

BULLETIN 43

The Geology of the Metamorphosed Biwabik Iron-Formation, Eastern Mesabi District, Minnesota

BY

JAMES NOVOTNY GUNDERSEN

AND

GEORGE M. SCHWARTZ

MINNEAPOLIS · 1962
THE UNIVERSITY OF MINNESOTA PRESS

PRINTED IN THE UNITED STATES OF AMERICA AT
THE JONES PRESS, INC., MINNEAPOLIS

Library of Congress Catalog Card Number: A62-9302

PUBLISHED IN GREAT BRITAIN, INDIA, AND PAKISTAN BY THE OXFORD UNIVERSITY PRESS
LONDON, BOMBAY, AND KARACHI, AND IN CANADA BY THOMAS ALLEN, LTD., TORONTO

FOREWORD

The construction of large concentrating plants and opening of pits for large-scale mining and milling of taconite during the past decade has emphasized the importance of all aspects of the geology of the Eastern Mesabi district.

It had been known from the earliest explorations in the Mesabi district that the eastern twenty-mile portion was characterized by a hard, siliceous, magnetite rock which Winchell called "taconyte." It was soon learned that this rock had not yielded to natural enrichment as had many areas in the main part of the range. Later, as the large complex concentrating plants went into operation, it became evident that intensive study of certain aspects of the geology, including detailed lateral as well as vertical stratigraphic variations, together with mineralogical and petrographic characteristics, would be extremely important to the successful operation of the huge pits and concentrating plants. The only detailed publication on the Eastern Mesabi was by Grout and Broderick in 1919, a study that necessarily depended mainly on outcrops, whereas large amounts of diamond drill core and extensive vertical exposures in the pits are now available.

The earlier mineralogical work was done before x-ray methods were well developed and before the complex amphibole and pyroxene groups were well understood. It was therefore obvious to the Director of the Minnesota Geological Survey that a modern detailed study was needed to supplement the excellent earlier work of Grout and Broderick. Fortunately, Dr. E. W. Davis, who had spent much of a lifetime on developing a process to concentrate taconite, was a consultant for the Reserve Mining Company. He fully understood the significance of a detailed knowledge of taconite, and as a result of his suggestions the company established an excellent postgraduate fellowship at the University of Minnesota to aid in fundamental research on the characteristics of taconite.

Dr. James N. Gundersen, currently of the Department of Geology, Los Angeles State College, was appointed to the fellowship. The Minnesota Geological Survey agreed to assume field and other expenses and to direct the work. The results published in this bulletin speak for the character of the work accomplished. The bulletin is adapted from Dr. Gundersen's thesis submitted for the degree of Doctor of Philosophy. He deserves the highest praise for the energy and devotion he has given to the problem.

GEORGE M. SCHWARTZ

ACKNOWLEDGMENTS

This project was supported in part by the Reserve Mining Company Research Fellowship, for the period July 1, 1955, to June 30, 1957. Gratitude is expressed for this aid as well as for the complete cooperation and assistance of members of the Reserve staff throughout the work. The Company generously drilled two additional holes through the entire formation at the request of the Minnesota Geological Survey to enable completion of the study. A separate grant was also made to provide for the chemical analyses reported herewith. Special thanks are due to Dr. E. W. Davis for his enthusiastic interest throughout the investigation. Mr. Raymond Berdie was largely responsible for company assistance throughout. The Erie Mining Company generously made available the core of a deep drill hole (No. 27E) and access to their open pits near Mesabi.

Field and laboratory expenses were assumed by the Minnesota Geological Survey. The analyses were made in the Rock Analysis Laboratory of the University of Minnesota, under the direction of Dr. S. S. Goldich, who also furnished advice on laboratory problems.

Thanks are due to Professor John W. Gruner and Dr. Hatten S. Yoder for stimulating discussions concerning the paragenesis of the metamorphic minerals of the Eastern Mesabi district. Dr. Gruner also furnished invaluable assistance in the verification and interpretation of x-ray studies of the minerals.

Appreciation is also extended to Professor S. R. B. Cooke for the use of facilities of the School of Mines and Metallurgy at the University of Minnesota and for generous interest in the problem.

Thanks are also due to Drs. James A. Whelan and Rolland L. Blake, who were working on related problems involving some of the ferrous silicates.

As in all extensive projects of the kind reported in this bulletin, many people not mentioned above rendered incidental aid which is sincerely appreciated. And finally, as will be evident to the reader, the early work of Grout and Broderick was of inestimable help.

CONTENTS

Foreword	v
Acknowledgments	vii
Abstract	xv
1. Introduction	1
General Statement	1
Location and Accessibility	1
History	3
Methods	4
Physical Features	5
Climate	5
Geologic Setting	5
Type Locality for Taconite Rock	6
Stratification Structure of Taconite	8
Classification of Taconite	8
2. Stratigraphy	11
General Statement	11
Giants Range Granite	16
Pokegama Formation	18
Contact between the Biwabik and Pokegama Formations	19
Biwabik Iron-Formation	22
General Features	22
Lower Cherty Member	24
Lower Slaty Member	31
Upper Cherty Member	35
Upper Slaty Member	55

Virginia Formation 68
Diabase Dikes and Sills 69
Duluth Gabbro 71
Pegmatitic Veins 72

3. MINERALOGY 76
Introduction 76
Elements 77
Sulfides 77
Arsenides 78
Oxides 78
Carbonates 81
Sulfates 81
Phosphates 82
Silicates 82
    Amphibole Group 82
    Feldspar Group 89
    Garnet Group 91
    Mica Group 91
    Pyroxene Group 92
    Other Silicates 95

4. METAMORPHISM 102
Introduction 102
Metamorphic Mineral Assemblages 102
Paragenetic Sequence of Metamorphic Minerals 104
The Iron Oxide-Silica-Water System 109
Interpretation of the Observed Petrologic System
    $FeO-SiO_2-H_2O-MgO$ 112
The "Primary" Silicates of the Biwabik Formation 118
Summary and Conclusions 121

| CONTENTS | xi |

5. PRACTICAL ASPECTS	124
General Discussion	124
Summary	133
REFERENCES	136
INDEX	137

LIST OF ILLUSTRATIONS

1. Aerial view of the Peter Mitchell Pit looking southwest	2
2. Index map for the Eastern Mesabi district, Minnesota	3
3. The five general taconite types	9
4. Typical drilling and loading operation in the Peter Mitchell Pit	14
5. Generalized columnar section of the Biwabik iron-formation	15
6. Contact of the Pokegama quartzite and the Biwabik iron-formation	20
7. Contact of the Pokegama quartzite and the Biwabik iron-formation	20
8. Contact of the Pokegama quartzite and the Biwabik iron-formation	22
9. Clastic fabrics in unit U	27
10. Almandine porphyroblasts of unit R	30
11. Granule structures replaced by stilpnomelane	30
12. Intraformational chert conglomerate of unit Q	32
13. Graphite-ferrohypersthene fabrics in unit Q	33
14. Ferrohypersthene remnants in amphibole matrix	40
15. Fayalite remnants in cummingtonite-rich matrix	41
16. Relic magnetite-quartz granules partly replaced by cummingtonite	43
17. Relic granule structures	44
18. Incipient development of cummingtonite	48

19. Riebeckite	51
20. Algal structures	52
21. Granule structures replaced by metasomatic hornblende	54
22. Relic granule structures	57
23. Shaly bedded taconite	59
24. Tabular cummingtonite	61
25. Laminated taconite	63
26. Relic bedded chert	64
27. Laminated taconite	66
28. Stilpnomelane	67
29. Relic granule structures in calcite	79
30. Hornblende+cummingtonite	88
31. Minnesotaite	97
32. Fayalite and ferrohypersthene	105
33. Granule-shaped structures in fayalite	106
34. Fayalite remnants in cummingtonite	107
35. Granule-shaped structures in cummingtonite	108
36. Petrologic system $FeO-SiO_2-H_2O$	110
37. Observed petrologic system $FeO-SiO_2-H_2O-MgO$	113
38. The generalized stratigraphy and magnetic concentration characteristics of taconite core from hole 21	126
39. The generalized stratigraphy and magnetic concentration characteristics of taconite core from hole 32	127
40. Grain fabrics in granule and shaly bedded magnetic taconites	130
41. Grain fabrics in laminated and layered magnetic taconites	131

INSERTED PLATE

Generalized geologic map and longitudinal diagram of the Eastern Mesabi district, Minnesota

LIST OF TABLES

1. Method of Classifying and Naming Taconites 10
2. Stratigraphic Succession and Geochronology of the Precambrian of Minnesota 12
3. Tabular Section of the Biwabik Formation on the Eastern Mesabi 13
4. Mineral Analyses of Samples from the Metamorphosed Iron Formation of the Eastern Mesabi District 76
5. Average Composition of Natural Taconite Feed 128

ABSTRACT

The economically important Eastern Mesabi district of Minnesota is the type locality for the iron-formation rock type *taconite*, a stratified quartzose rock containing significant amounts of iron-bearing oxides, hydroxides, silicates, and, locally west of the district, carbonates. Five basic types of taconite — massive, layered, laminated, shaly bedded, and shaly — are delineated for detailed classification of the stratified structure and mineralogy of the Biwabik iron-formation in this district.

The previous stratigraphic division of the Precambrian Biwabik formation into five members has been further divided into 22 submembers that can be recognized nearly throughout the region. The delineation of these individual units is based on the relative abundance of the basic taconite types, mineral assemblages, and the presence of relics of primary sedimentary structures such as algal remains and conglomeratic, brecciated, and granule-rich zones. The lateral variations of mineral assemblages within each of the stratigraphic units provide the background for the discussion of the metamorphic history of the formation.

Although quartz and magnetite dominate the mineralogy of the taconite, there is also a significant number of metamorphic silicates in the district. Among these are cummingtonite, fayalite, ferrohypersthene, hedenbergite, hornblende, actinolite and to a lesser extent andradite, diopside, biotite, hisingerite, riebeckite, and stilpnomelane. Traces of several other silicates, especially feldspar, occur locally. Small amounts of pyrite, pyrrhotite, and loellingite are also present.

Considerable field and petrographic evidence exists to indicate that the constituents of quartz and magnetite are recombined to form most of the metamorphic and metasomatic mineral assemblages observed. There is, on the other hand, no evidence indicating the previous existence of the so-called primary silicates or carbonates of the iron formation nor any evidence supporting the concept of progressive metamorphism of these minerals into any of the mineral assemblages now found in the district. The theoretical implications of these observations and the proposal of original quartz-magnetite assemblages are discussed.

The emplacement of the Duluth gabbro resulted in thermal metamorphism of the adjacent rocks, which was manifested largely by simple recrystallization and formation of fayalite-magnetite-quartz assemblages in the iron formation. Almost contemporaneously, the injection of numerous pegmatite-like veins was accompanied by the introduction of magnesium- and calcium-bearing aqueous solutions into the adjacent taconites, which resulted in the widespread formation of cummingtonite, hedenbergite, and some calcium-bearing ferrous amphiboles. Local arsenide and sulfide mineralization was also associated with this metasomatic activity.

The magnetite from some of the stratigraphic intervals is currently being extracted for the manufacture of iron-oxide pellet feed for blast furnaces. By means of a few examples, the importance of the primary bedding structures and mineral textures of each of the taconite types as affecting the quality and quantity of a magnetic concentrate is stressed. The probable magnetic concentration characteristics of each stratigraphic submember are discussed in the light of its mineralogy and grain textures, and the modifying effects due to possible metamorphic changes in mineralogy and grain fabric are also emphasized.

THE GEOLOGY OF THE
METAMORPHOSED BIWABIK IRON-FORMATION, EASTERN MESABI DISTRICT, MINNESOTA

1. INTRODUCTION

GENERAL STATEMENT

Future large-scale mining operations of the Reserve Mining Company on the Eastern Mesabi district will eventually result in a practically continuous pit for a distance of fully eight miles, with a generally southwesterly trend from the initial opening near Old Babbitt (Fig. 1). Inasmuch as the huge Duluth gabbro intrusion transgresses the Biwabik iron-formation to the east, it has long been known that metamorphism is more intense near the contact (Leith, 1903, p. 159). This, in turn, means that changes in the character of the formation occur along the length of the range, particularly in the eastern portion. These changes are of great significance in the concentration process where requirements for a high iron and low silica content are strict. It is therefore desirable, or indeed necessary, to have as much information as possible regarding the detailed stratigraphic and mineralogical relations of the Eastern Mesabi district.

LOCATION AND ACCESSIBILITY

The Mesabi district, or range as it is commonly called, extends from near Remer west of Grand Rapids eastward to Birch Lake, a distance of about 128 miles. This is from longitude 47° 30' and latitude 93° 52' to longitude 47° 45' and latitude 91° 50'. The distance along the strike of the formations is somewhat greater owing to the broad fold in the Virginia-Eveleth area usually referred to as the Virginia Horn.

The Eastern Mesabi district refers to that part of the district extending from near the village of Mesaba to just beyond Old Babbitt, a distance of about 19 miles (Fig. 2). At Birch Lake the Duluth gabbro cuts off or covers the iron formation. The Eastern Mesabi district is essentially 55 miles due north of Duluth.

This bulletin is mainly concerned with the part of the district lying east of rangeline 14 west, that is, land lying in townships 59, 60, and 61 north, ranges 12 and 13 west. In the discussions that follow, such phrases as "the area studied" or the "Eastern Mesabi range" apply to that part of the district.

The detailed work showed that in R. 14 W., and farther west, the minerals of the iron formation (taconite) grade into the low grade metamorphic assemblages that are more typical of the main range (Gruner, 1946).

The Eastern Mesabi district is served by one common carrier railroad, the Duluth, Mesabi and Iron Range, with lines to Duluth and

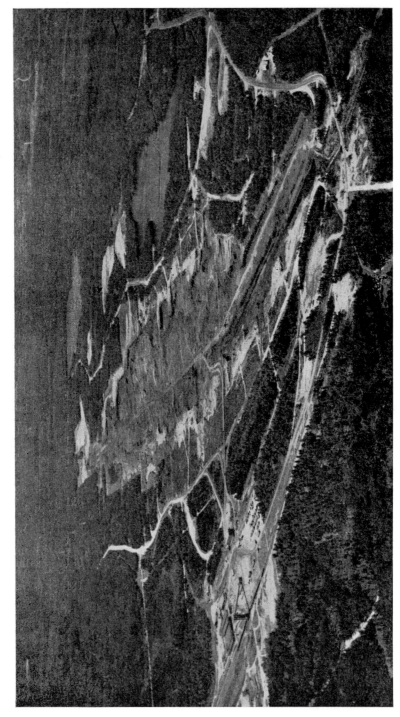

FIGURE 1. — Aerial view of the Peter Mitchell Pit looking southwest. Argo Lake is to the upper right, Iron Lake beyond. The pit is about three and one-half miles long, but will eventually extend southwest an additional five miles. (Photo courtesy Reserve Mining Company, June 13, 1959.)

INTRODUCTION

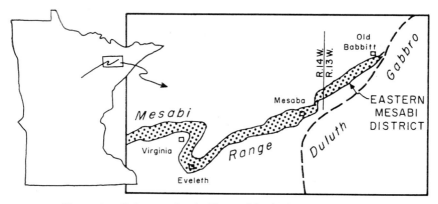

FIGURE 2. — Index map for the Eastern Mesabi district, Minnesota.

Two Harbors. Two private railroads serve the Reserve and Erie properties with terminals on Lake Superior at Silver Bay and Taconite Harbor respectively.

HISTORY

The earliest references to iron-bearing rocks of the Mesabi district were concerned with the abundant outcrops of the eastern part of the district lying between Embarrass Lake (now drained) and Birch Lake. It is within this area that the two larger taconite mining operations of the Erie and Reserve Mining companies are now centered. To the west the iron-bearing rocks are largely covered by glacial drift.

The earliest detailed recorded examination of this area by a mining expert was made in 1875 by A. H. Chester (1884, p. 154–167), who noted the magnetic effect of the rocks but considered them too low grade for mining, which was certainly true at the time. Earlier, in 1866, however, Henry H. Eames, the first state geologist of Minnesota, referred to the elevated area of the northern part of the state and noted that enormous bodies of iron ore occurred, but the observations appear to have been made at the west end of the range.

According to Chester's report (1884, p. 157) the first explorer of the Mesabi Range was Peter Mitchell, who excavated a pit six feet in depth in the northwest quarter of Sec. 20, T. 60, R. 12 W. in 1871. The site is near the present large crushing plant of the Reserve Mining Company, and the present open pit mine is appropriately named the Peter Mitchell Mine.

More detailed observations in the eastern part of the district, now generally referred to as the Eastern Mesabi, were made by N. H. Winchell in 1879 and 1881. In the 7th Annual Report of the Geological and Natural History Survey, N. H. Winchell referred to a costly exploration in Ts. 59 and 60, R. 14, a part of the East Mesabi district. He noted the magnetic character of the rocks and included a number of analyses made by prospectors who obviously must have taken their samples from a selected magnetite bed to yield the high iron content cited.

The 9th Annual Report presented some details regarding exploration pits, rocks, and ores of the eastern part of the Mesabi Range, and in the 17th Annual Report H. V. Winchell described a traverse along the range from Birch Lake to the Duluth and Iron Range railroad, covering essentially the Eastern Mesabi district as now known. He also emphasized the abundance of magnetite. The Annual Reports from 1889 on contain various observations on the district.

Bulletin 6 by N. H. and H. V. Winchell, published in 1891, contains a brief reference to the discovery of hematite ores near Mountain Iron on November 16, 1890. This served to divert attention from the low grade magnetite-bearing rocks of the Eastern Mesabi for nearly three decades, although sporadic attempts were made to discover workable deposits.

The beginning of modern developments was in 1915 when the Mesabi Syndicate was organized, with D. C. Jackling taking an important part. In 1916 an experimental plant was built in Duluth and a concentration process developed. The Mesabi Iron Company was incorporated in Delaware, December 12, 1919, and took over from the Syndicate. Construction of a plant was started at Babbitt, on the range, and was completed in July 1922. For several reasons the plant was unsuccessful and was closed in June 1924.

As time passed it dawned on some of the forward-looking men in the steel industry that the high grade ores were not going to last indefinitely and steps were taken to investigate large resources of low grade but concentratable material. Attention was inevitably drawn to the Eastern Mesabi district, and in 1939 the Reserve Mining Company was organized by Oglebay-Norton & Company and leases were obtained from the Mesabi Iron Company.

Experimental work on the beneficiation of taconite continued at the Mines Experiment Station, University of Minnesota, under E. W. Davis, but not until 1947 was a decision made by the Reserve Mining Company to erect a large concentrating plant. The result is Reserve's large mine and the crushing plant on the range at Babbitt, a railroad from Babbitt to Silver Bay on Lake Superior, and the large concentrating plant at Silver Bay.

Meanwhile other steel companies became interested in the Eastern Mesabi district. As a result, the Erie Mining Company was formed to mine taconite from lands lying west of those controlled by Reserve. This plant went into operation in 1958.

METHODS

During the summers of 1955, 1956, and part of 1957, an intensive study of diamond drill cores and examinations of artificial exposures in the Mitchell pit were completed. The Erie Mining Company made available an especially valuable set of cores from a drill hole in their area west of the Reserve property. Since 1957 additional field work has been continued sporadically, on a small scale. Extensive laboratory studies beginning in the Fall of 1955 were carried on until 1958 and have continued less intensively since that time. Modern mineralogic and petrographic methods have been utilized as well as chemical analysis.

It was found early in the investigation that it was unnecessary to attempt remapping of outcrops because of the excellence of Grout and Broderick's map (1919) and also because a heavy second growth of brush since that time had covered many outcrops, or moss and lichens had masked their features.

Necessary comparisons were made with the stratigraphy and mineralogy of the main range where appropriate.

PHYSICAL FEATURES

The common term of range used for the Mesabi district is derived from the long high ridge formed primarily by the Giants Range granite with the Biwabik and adjacent formations lying on the south slope. The topography of the Eastern Mesabi district is shown on the Allen, Aurora, Babbitt, Babbitt N.E., and Aurora 7½′ topographic sheets published by the U.S. Geological Survey. Immediately south of the iron formation is a large swamp named by the early explorers the "One Hundred Mile Swamp." This is bordered on the North by the 1620-foot contour. To the north the altitude increases irregularly to the crest of the granite ridge where the highest parts are generally above altitude 1800 but reaching 1940 in the so-called Embarrass Mountains northwest of Mesaba. Northeast of the old townsite of Babbitt, the ridge dies out rapidly near the Dunka River and descends to 1420 at Birch Lake. The new townsite of Babbitt is located on a level area north of the ridge at about 1480 and less than a mile south of Birch Lake.

CLIMATE

The location of the Mesabi district north of Lake Superior and at a relatively high altitude for the Precambrian Shield results in heavy snowfall and generally severe winters. The annual precipitation is about 30 inches with an average of approximately 60 inches of snow. The minimum temperatures approach 50° F below zero and the maximum 100° F above. The January mean is about 7°, that for July about 64°, and the mean annual about 37°.

GEOLOGIC SETTING

The Precambrian rocks of Northeastern Minnesota form a complex group that includes what have long been considered classic examples of many aspects of Precambrian geology. Accordingly these rocks have received intensive study by the Minnesota Geological Survey, the United States Geological Survey, and by geologists of the various mining companies and universities throughout the Middle West. The chronologic sequence is given in Table 2.

The formations of the Mesabi Range include small areas of the oldest rocks; that is the Ely greenstone and Knife Lake slates and graywackes which were intruded by the Giants Range granite. The resistant granite results in the characteristic ridge north of the iron-bearing rocks.

The oldest formation is the Ely greenstone, a group of basic volcanic rocks which have been highly altered by various metamorphic processes including a considerable amount of hydrothermal activity. Chlorite has been formed by alteration of the primary ferromagnesian minerals, hence the green color.

The main areas of greenstone lie well north of the Mesabi district, but small areas lie at the north edge of the iron formation near Chisholm, Mountain Iron, and a larger mass in the Virginia Horn area.

The Knife Lake group overlies the greenstone and is also represented by a small area between the Giants Range granite and the Animikie group, consisting of slate, graywacke, and conglomerate.

For the most part the Animikie rocks lie unconformably on the Giants Range granite, which extends the full length of the range and somewhat beyond both to the east and west. The known length of the Biwabik formation is about 105 miles. At the east end near Birch Lake the formation is overlain and transgressed by the Duluth gabbro, but appears again in the Gunflint district fifty miles to the east.

The long narrow extent of the Biwabik formation is a result of a southeasterly dip of 5 to 15 degrees related to its position on the north side of the Lake Superior syncline.

The Eastern Mesabi district comprises the part of the Mesabi range about 20 miles from Birch Lake west to within a mile of the town of Mesaba. This part was directly affected by the intrusion of the Duluth gabbro, and for that reason resisted the leaching that formed the high grade ores to the west.

TYPE LOCALITY FOR TACONITE ROCK

Taconite has been used as a rock name to describe the rocks of numerous stratified Precambrian sedimentary iron-formations of the world, but the Eastern Mesabi district is the type locality for the rock type "taconite." The rock name has proven so useful, however, that it would be senseless to restrict the name to the iron-formation rocks on the Eastern Mesabi range. The writers agree that the manner in which Gruner (1946, p. 5) has employed the term seems best suited to name the quartzose rocks of stratified Precambrian iron-formations that contain significant amounts of iron-bearing silicates, carbonates, oxides, and hydroxides without restrictions as to whether the rocks are sediments or metasediments, or if they are fresh or partially oxidized — with the exception of the oxidized and enriched ores.

Although the Biwabik iron-formation in the Eastern Mesabi district has been substantially metamorphosed, its relic sedimentary stratification is apparent almost everywhere, a feature which greatly facilitates lithologic classification of these rocks. Most of the rocks of the iron formation throughout the entire Mesabi range consist largely of four main mineral groups: (1) quartz (and chert); (2) iron-rich silicates; (3) iron-bearing carbonates; and (4) iron oxides and hydroxides. On the Main Mesabi range most of the silicates are minnesotaite, stilpnomelane, and greenalite, all of which are difficult to recognize in hand specimen because of their fine grain size. The gross mineralogy of the taconites from the Eastern Mesabi range is simple, consisting essentially of magnetite, quartz (including chert), and iron-rich silicates. For a general classification of taconite types, the silicates can be considered as a group, but where the silicates can be recognized the appropriate mineral name should be used in naming the specific taconite rock type. The most abundant silicates found at the type locality are cummingtonite, ferrohypersthene, fayalite, and lesser amounts of actinolite, hornblende, and hedenbergite, all of which are commonly sufficiently coarse-grained to be recognized with the hand lens. There seems to be little doubt that nearly all the quartzose layers of the iron formation once consisted of fine-grained silica, probably chert. In a rigorous sense, however, chert is rare in the Eastern Mesabi range. Therefore, as a compromise to avoid losing the use of two useful mineral names, bedded silica that is easily resolvable with a 10X hand lens as distinct grains (mostly about 0.1 mm in diameter or larger) is called quartz, and the more finely crystalline or microcrystalline quartzose material is called chert.

The results of beneficiation testing indicate that the concentrating characteristics of the magnetic taconites are closely related, in part, to their original bedding features (see Chapter 5). The wide variety of taconite types of somewhat similar mineralogy, and relic sedimentary structures found in the magnetic taconites, made it apparent that a detailed but workable lithologic classification of the Eastern Mesabi taconites could be helpful in predicting the beneficiation characteristics of the various taconite types, as well as in the recognition and unique description of the numerous stratigraphic units of the iron formation. The classification of Mesabi range taconites in general has been discussed by Gruner (1946, p. 32) and by White (1954, p. 7). Gruner's classification is simple and easy to apply, but the lack of a specific notation or system for noting the mineral assemblages actually occurring in the various rock layers of a given taconite type makes his scheme too inclusive for obtaining detailed rock names. White's classification, based largely on the presence or absence of granule structures, is more complicated and is difficult to apply to the detailed study of magnetic taconites because it has been experienced that features found in thin section can change the basic rock name arrived at by an examination of the hand specimen.

A detailed lithologic classification of the taconites from the type locality has been proposed (Gundersen, 1960a) which is essentially an expanded but more specific version of Gruner's scheme. It is based upon the features actually present in the rock and not upon the absence of certain features. It is consistent regardless of the mode of examination of the rock, so that additional information obtained by microscopic study will merely augment the rock name but will never change the basic taconite name.

STRATIFICATION STRUCTURE OF TACONITE

The Biwabik iron-formation in the Eastern Mesabi district consists largely of two distinctly different kinds of strata, herein defined as *bedded taconite-strata* and *massive taconite-strata*, that are interbedded to collectively form almost all of the iron formation. Both kinds of taconite-strata are schematically illustrated in Figure 3. Although both types of strata are bedded in the true sense of the word, bedded taconite-strata includes the layers or beds that are responsible for the obvious layered or stratified aspect of the formation. These beds generally contain most of the iron oxides of the formation, but there are important exceptions. Most or all of the "bands" and many of the "slaty beds" of older classifications correspond to taconite-strata of this type. In contrast, massive taconite-strata are the layers or beds that are massive in the sense that they are normally devoid of any internal stratification. These taconite-strata correspond to the massive layers of "silicate taconite" and "cherty taconite" of existing schemes. Both types of taconite-strata range in thickness from almost zero to several feet, but the great majority of individual layers are less than 6 inches.

In addition to their differing mineral assemblages, the most apparent difference between the taconite rock types of the iron formation is the average thickness and abundance of one kind of taconite-strata relative to the other. This significant feature is the ultimate basis for naming the general types of taconite in the Eastern Mesabi district.

CLASSIFICATION OF TACONITE

Five general types of taconite, each based upon the relative abundance of bedded taconite-strata that are interbedded with massive taconite-strata, are proposed in the classification used for naming the rocks of the iron formation.

These five types are defined as follows:

1. *Massive taconite*, consisting almost entirely of massive taconite-strata with from zero to about 10 per cent by volume of bedded taconite-strata.

2. *Layered taconite*, consisting mainly of massive taconite-strata interbedded with about 10 to 40 per cent by volume of bedded taconite-strata.

3. *Laminated taconite*, containing about 40 to 60 per cent by volume

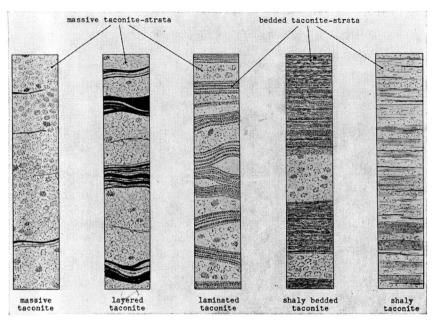

FIGURE 3. — The five general taconite types. Simplified sketches showing generalized distribution of bedded and massive taconite-strata.

of bedded taconite-strata with the remainder consisting of massive taconite-strata.

4. *Shaly bedded taconite,* containing about 60 to 90 per cent by volume of bedded taconite-strata with subordinate amounts of massive taconite-strata.

5. *Shaly taconite,* consisting almost entirely of bedded taconite-strata with from zero to about 10 per cent by volume of massive taconite-strata.

Words connoting structural features are used as prefixes to the general rock names, as has been done in other taconite classifications, e.g., "banded taconite." The structural prefixes selected seem to best distinguish between the over-all bedding aspects of the rocks in question. To be sure, these words are not mutually exclusive, because, for example, the shaly bedded rocks are certainly both laminated and layered. Consequently, the structural prefixes to the rock names must be recognized only as convenient comparative prefixes used to designate a particular type of taconite.

The next step in this classification is to have a scheme by which the mineral assemblages of both the bedded taconite-strata and the massive taconite-strata can be included in the rock name. In brief, the mineralogy of the massive taconite-strata determines the basic rock name of the specific taconite in question. The mineralogy of the bedded taconite-

TABLE 1. METHOD OF CLASSIFYING AND NAMING TACONITES

General Taconite Types	Volume Per Cent of Bedded Taconite-Strata	Specific Taconite Rock Names
Massive	0 to 10	C–D taconite
Layered	10 to 40	layered (A–B) C–D taconite
Laminated	40 to 60	laminated (A–B) C–D taconite
Shaly bedded	60 to 90	shaly bedded (A–B) C–D taconite
Shaly	90 to 100	shaly A–B taconite

strata is also incorporated into the rock name by setting it off in parentheses immediately following the structural prefix portion of the rock name. For example, a type of layered taconite consisting of magnetite layers in a silicate-bearing quartzose matrix is given the rock name *layered (magnetite) silicate-quartz taconite*. The general rule is that in listing the mineralogy of both the bedded and massive taconite-strata, the least abundant mineral is written first and the most abundant last. Other structural prefixes are used to a lesser extent to indicate the presence of other textures and structures found particularly within the massive taconites — for example, mottled grain fabrics, conglomeratic zones, and abundant granule structures, among others. Most of the structural prefixes used are obvious, but further description is available (Gundersen, 1960a) for the interested reader. Again, the minerals that constitute these structures are set off in parentheses.

In naming massive taconites, the prefix *massive* is not used and the over-all mineralogy of the taconite is sufficient to name the rock, because of the general absence of bedded structures. Type 5, shaly taconite, is perhaps best visualized as a special variety of silicate-rich massive taconite that has a distinct thin-bedded aspect. In these rocks the segregation of silicates, quartz, and, locally, graphitic material into somewhat concentrated, yet diffuse, regularly-bedded lamellae, thin layers, and partings is sufficient to produce the over-all shaly aspect throughout the rock. Because of the lack of abundant, distinct layers that are appreciably different mineralogically from their matrix, the usual system of placing their dominant mineralogy in parentheses is not applicable. Consequently, rocks of this type are prefixed with the word *shaly* and the over-all mineralogy of the taconite is used to name the rock.

In all of the rock names given above, it has been assumed that the stratification is largely regularly bedded. Many zones within the stratified taconites, however, are irregularly bedded locally and nonuniform; the additional prefix *wavy* has been used to designate this feature.

The general scheme of the classification used is summarized in Table 1. Without being specific, assuming that the mineralogy of the massive taconite-strata consists dominantly of mineral D and lesser amounts of mineral C, and similarly, that the bedded taconite-strata consist mainly of mineral B and lesser amounts of mineral A, then Table 1 also shows the simple scheme used in naming the specific taconite types.

2. STRATIGRAPHY

GENERAL STATEMENT

The stratigraphic position and stratigraphy of the Biwabik iron-formation have been discussed in detail many times by previous authors, especially Wolff (1917), Gruner (1924, 1946), Grout, et al. (1951), and White (1954), and the reader is referred to these reports for discussions of all features relating to the Mesabi range in general.

The chronologic and stratigraphic sequence in northern Minnesota is best summarized in Table 2, as adapted from Goldich, et al. (1961). Of the many stratigraphic units listed in this table, only the Biwabik formation as it occurs in the Eastern Mesabi district, and the pegmatitic veins that cut it here, are discussed in detail. The remaining sedimentary units of the Animikie group, the underlying Virginia formation and the underlying Pokegama formation, as well as the intrusions of the area, the Giants Range granite, diabase sills and dikes, and the Duluth gabbro, are described in less detail.

The most important publication concerning the Eastern Mesabi district is Bulletin 17 of the Minnesota Geological Survey, *The Magnetite Deposits of the Eastern Mesabi Range, Minnesota* by F. F. Grout and T. M. Broderick (1919). These authors made a detailed surface map of the entire area in 1917 shortly after an extensive forest fire revealed excellent exposures of the formation. Thick undergrowth of the regrown forest has effectively prevented the writers from doing additional surface mapping. Because of the close agreement and easy correlation between the stratigraphic units of Grout and Broderick and the units proposed herein, as well as the ideal conditions under which these previous investigators worked, it is believed that the earlier geologic map could not be effectively improved upon by additional field mapping.

The stratigraphic column, as proposed by Grout and Broderick (1919, p. 17) and Broderick (1919, p. 444), is given in Table 3. They recognized, of course, that the Slaty and Cherty members of the Main Mesabi range were not recognizable as such, but the probable correlation of these units to their own is indicated in their stratigraphic section.

The present writers had very distinct advantages over the previous workers because of the availability of many recent drill cores and extensive exposures (Fig. 4) of the Upper Cherty member of the iron formation as a result of the current mining operations of the Reserve Mining Company. In addition, the Reserve Mining Company generously provided two new drill holes (numbers 21 and 32) that yielded core from the entire Biwabik formation, and part of the Pokegama forma-

TABLE 2. STRATIGRAPHIC SUCCESSION AND GEOCHRONOLOGY OF THE PRECAMBRIAN OF MINNESOTA

Era (10⁹ Years)	Period-System	Major Sequence	Formation	Orogeny	Intrusive Rocks
Paleozoic (0.6 b.y.)	Cambrian		............Unconformity............		
Late Precambrian (1.1 b.y.)	Keweenawan		Hinckley sandstone		
			Fond du Lac sandstone		
			............Unconformity............		
		North Shore volcanic group	Undivided	Grenville	Duluth complex, sills at Duluth, Beaver Bay complex, Logan intrusives
			Puckwunge		
			............Unconformity............		
			Sioux quartzite (?)		
(1.7 b.y.)			............Unconformity............	............Penokean	Granite: St. Cloud Red, Rockville (?) granite at Granite Falls, Bellingham (?)
			Virginia argillite = Rove = Thomson		Gneiss: McGrath, Montevideo (?)
		Animikie group	Biwabik iron-formation = Gunflint		Tonalites: St. Cloud Gray, Warman, Hillman, Freedhem, Montevideo
Middle Precambrian	Huronian		Pokegama quartzite		
			............Unconformity............	............Algoman	Granite: Gold Island, Giants Range, Sacred Heart, Fort Ridgely (?)
(2.5 b.y.)					Gneiss: Giants Range, Vermilion, Morton
(? b.y.)	Timiskamian	Knife Lake group	Undivided		
			............Unconformity............	............Laurentian	Saganaga granite, Grassy Island tonalite (?)
			Soudan iron-formation		
Early Precambrian	Ontarian	Keewatin group	Ely greenstone		
		Coutchiching (?)	Undivided		
			Older rocks		

12

STRATIGRAPHY

TABLE 3. TABULAR SECTION OF THE BIWABIK FORMATION ON THE EASTERN MESABI
(after Grout and Broderick, 1919)

VIRGINIA FORMATION

BIWABIK IRON-FORMATION

		Thickness (feet)
Upper Slaty Member		
Aub_6	Limy carbonate layers. Less than 5 per cent magnetite	5 to 10
	Banded amphibole and white quartz in thin layers up to 6 inches thick. Less than 5 per cent magnetite	40 to 50
Aub_5	Taconite in thin beds mostly less than 1/8 inch thick. About 20 per cent iron in magnetite, decreasing at top	25 to 35
	Taconite in thin beds, like above, but alternating with thicker gray beds and concretions with a granule texture, and white quartz septaria. The thin beds are drag-folded and even brecciated. A zone of garnets occurs near the bottom. About 20 per cent iron in magnetite	40 to 45
Upper Cherty Member		
Aub_4	Dark heavy taconite, with some conglomerate texture, and granules, alternating in thick beds with thick magnetite layers. About 30 per cent iron in magnetite	10
	Jasper and chert with algal structure and conglomerate. About 30 per cent iron in magnetite	1 to 10
	Dark heavy taconite like the bed above the jasper. About 30 per cent iron in magnetite	20 to 30
	Gray taconite with conglomerate and granule texture and many thinner lenticular beds of magnetite. About 20 per cent iron in magnetite	80 to 100
Lower Slaty Member		
Aub_3	Fine massive to slaty quartz amphibolite with only obscure granule structure, and few magnetite beds. Fayalite crystals in places. About 10 per cent iron in magnetite	65
Aub_2	Black thin-bedded slate, more or less recrystallized. About 5 per cent iron in magnetite	25
Lower Cherty Member		
Aub_1	White to gray chert with coarse algal structures. Less than 5 per cent iron in magnetite	10 to 15
	Variable taconite with some cherts, breccias, fragmental sands, garnets, etc. From 25 to 35 per cent iron in magnetite	2 to 52
	Basal beds; conglomerate in many places, and (in absence of Pokegama quartzite) some green shales. Less than 5 per cent iron in magnetite	0 to 15
		350 to 470

POKEGAMA FORMATION OR GIANTS RANGE GRANITE

tion, specifically for the present study. The cores from these new holes, and from over thirty previous holes, provided most of the data for the compilation of a new detailed stratigraphic column for the Eastern Mesabi district. The writers were also fortunate in being able to study excellent core from drill hole 27E, near Mesaba, of the Erie Mining Company as well as to examine the Upper and Lower Cherty members of the formation in that company's active open pits west of R. 14 W. The locations of all drill holes are indicated on the generalized geologic map accompanying this report.

FIGURE 4. — Typical drilling and loading operation in the Peter Mitchell Pit. The blocky nature of the taconite is typical. (Photo courtesy Reserve Mining Company.)

The Biwabik iron-formation on the Eastern Mesabi range was subdivided into many submembers (Gundersen 1958, 1960b), some of which differ only slightly from one another. It is not intended that the submembers described below should serve as a substitute for the units proposed by Grout and Broderick or the more commonly used Slaty and Cherty member terminology. The proposed column is in a sense academic, in that some submembers that can be recognized as distinct lithologic units in drill core are commonly difficult to differentiate in outcrop or pit exposures. Strictly speaking, the submembers are vertically segregated lithostratigraphic units with arbitrary horizontal cutoffs, and the lithology of each submember can be recognized by its homogeneity or, as in some submembers, by its characteristic heterogeneity. Most of the contacts between the submembers are gradational and hence have been located subjectively. The lateral variations in mineralogy and mineralogical assemblages within the iron formation submembers are interpreted to be largely a result of metamorphism and metasomatism, and not the result of variations of the initial composition within submembers. Hence, no laterally segregated lithostratigraphic units with arbitrary vertical cutoffs (primary sedimentary facies) can be proposed.

Members (Main Mesabi) (Wolff, 1917)	Members (Eastern Mesabi) (Grout-Broderick, 1919)	Submembers (Eastern Mesabi) (Gundersen, 1958)	& (approx thick, feet)	GENERALIZED COLUMNAR SECTION OF THE BIWABIK IRON-FORMATION IN THE EASTERN MESABI DISTRICT, MINNESOTA Description of Submembers (Notations (east) and (west) refer to areas near the eastern and western drill holes)
UPPER SLATY	Aub$_6$	A	(5)	calcite marble; minor diopside, wollastonite, idocrase, andradite and quartz
		B	(16)	layered (diopside) chert taconite locally with hornblende, hedenbergite and some cummingtonite and actinolite
	Aub$_5$	C	(42)	laminated (ferrohypersthene-magnetite) quartz taconite with hedenbergite and fayalite (east) and laminated (cummingtonite-magnetite) chert taconite (west)
				wavy laminated (actinolite-magnetite) chert taconite locally with cummingtonite and minor hedenbergite
		D	(7)	quartz taconite with abundant granule structures and locally with quartz-filled septaria structures; minor magnetite, cummingtonite and actinolite
		E	(6)	
		F	(20)	shaly bedded (cummingtonite-magnetite) quartz taconite with minor andradite and hedenbergite (east); locally abundant cummingtonite (west)
		G	(25)	quartz taconite (east) and mottled (andradite) quartz taconite (west) with abundant magnetite-bearing granules throughout
UPPER CHERTY	Aub$_4$	H	(10)	wavy layered (actinolite-magnetite) quartz taconite with minor hedenbergite; locally with fayalite (east) and cummingtonite (west)
		I	(5)	algal (magnetite) quartz taconite with abundant magnetite-rich granules and pebbles; conglomeratic fabric throughout; minor hematite
		J	(16)	granule (magnetite) quartz taconite with abundant magnetite-rich pebbles near top and thickly layered (magnetite) quartz taconite near bottom
		K	(35)	wavy layered (silicate-magnetite) quartz taconite with abundant magnetite-rich granules and pebbles; silicates are actinolite and ferrohypersthene (east) and cummingtonite (west)
		L	(30)	wavy layered (silicate-magnetite) silicate-quartz taconite with abundant magnetite-rich granules near bottom; silicates with magnetite are ferrohypersthene and hornblende (east) and cummingtonite and actinolite (west); silicates with quartz are ferrohypersthene (east) and cummingtonite (west)
		M	(20)	layered (magnetite) fayalite-quartz taconite with ferrohypersthene (east) and layered (magnetite) cummingtonite-quartz taconite (west)
		N	(4)	fayalite-quartz taconite with ferrohypersthene (east) and cummingtonite-quartz taconite (west); minor magnetite
		O	(17)	bedded granule (magnetite) quartz-fayalite taconite with some ferrohypersthene and minor cummingtonite (east) to quartz-cummingtonite taconite with magnetite-bearing granules (west)
LOWER SLATY	Aub$_3$	P	(60)	shaly quartz-fayalite taconite to fayalite taconite (east) and shaly quartz-cummingtonite taconite to cummingtonite taconite (west); minor magnetite
				argillaceous graphite-silicate-quartz taconite with abundant ferrohypersthene and minor fayalite, biotite, almandite and pyrrhotite (east) and traces of pyrite, pyrrhotite and cummingtonite (west)
				layered (magnetite) fayalite-quartz taconite with minor cummingtonite
	Aub$_2$	Q	(26)	layered (magnetite) quartz taconite with minor cummingtonite throughout, and hedenbergite and some fayalite (east)
				granule (magnetite) quartz taconite with minor cummingtonite throughout, and minor fayalite (east)
LOWER CHERTY	Aub$_1$	R	(11)	layered and granule (magnetite) cummingtonite-quartz taconite with hedenbergite locally, and some fayalite (east)
		S	(8)	
		T	(5)	
		U	(10)	quartz taconite with minor hedenbergite and cummingtonite; clastic quartz pebble zone locally at base
		V	(3)	

FIGURE 5. — Generalized columnar section of the Biwabik iron-formation.

Although the resultant stratigraphic column contains many more lithostratigraphic units (submembers) than the earlier column proposed by Grout and Broderick, the correlation with their units is easily made. Figure 5 shows the approximate correlation between the stratigraphic units proposed by Wolff (1917), Grout and Broderick (1919), and those recognized by Gundersen (1958, 1960b).

By convention the thicknesses of stratigraphic units and individual beds are described in units of feet and inches, while the sizes of mineral grains and other microscopic features are presented in millimeters and centimeters.

GIANTS RANGE GRANITE

The Giants Range Granite was encountered in core from holes 17, 26, 32, and 34; many natural outcrops occur along the northern outcrop limit of the Biwabik formation, particularly near Old Babbitt townsite.

In the natural outcrops the rock is a pink-white, porphyritic (hornblende) biotite granite. Euhedral to subhedral, pink silicic feldspar (cloudy orthoclase?) occurs as phenocrysts that commonly exceed 10 mm in diameter. The subhedral plagioclase laths of the groundmass are clear to milky white but are locally tinted pink because of admixtures of silicic feldspar. Neither of the feldspars shows sharp grain boundaries, and seems to pass somewhat indistinctly into a fine-grained, biotite-rich quartzose matrix. Much of the biotite is segregated into clusters that are mainly interstitial to the feldspar grains; in numerous places, however, thin streaks of biotite almost completely surround the smaller (1 to 3 mm diameter) grains of both feldspars. In some places the biotite is segregated into veinlet-like streaks and small schlieren. In general, this description applies equally well to the core specimens of the granite, with the exception that in some places the rock is distinctly less porphyritic and that a few cores show pronounced gneissic foliation.

The specimens of the granite examined in thin section do not show simple igneous textures but have fabrics that might be more closely associated with hybrid or partially metasomatized igneous rocks. The orthoclase phenocrysts are commonly clouded, subhedral in shape, with their grain boundaries irregular in detail. The plagioclase grains (about An_{15}), along with quartz and orthoclase, locally occur in a granoblastic-like mosaic which generally shows irregular to nearly sutured grain boundaries. In some places the plagioclase occurs as laths poikilitically enclosed within the orthoclase phenocrysts. In this fabric the enclosed plagioclase laths contain cloudy centers and thin, clear overgrowths, but the grains do not show primary compositional zoning. The groundmass immediately surrounding these orthoclase phenocrysts contains similar plagioclase laths with thick, clear overgrowths in a groundmass of anhedral orthoclase and quartz grains. Both feldspars have locally been altered to a cloudy white material (kaolin?) and to white mica (sericite?). In some specimens the cores of the plagioclase laths appear to be slightly saussuritized; in thin section, however, these grains are seen to have been partly replaced instead by green (and some brown) biotite and secondary chlorite.

The mafic minerals of the granite are generally segregated into numerous small clusters and streaks of grains that are mainly interstitial to the feldspars and quartz. Green and brown biotite, commonly occurring in nearly equal amounts, are the most common mafic minerals in the granite. Both biotites commonly contain strong pleochroic haloes surrounding minute, radioactive zircon grains, and both have been partly altered to chlorite (penninite) in most places. Roundish clusters of biotite and chlorite are sometimes found poikilitically enclosed within orthoclase phenocrysts. Green hornblende occurs to a minor extent and exists in some places only as remnants within clusters of biotite. The mafic minerals are commonly associated with accessory amounts of zircon, apatite, sphene, magnetite, and ilmenite (with some secondary leucoxene).

Examination of the granite outcrop areas, particularly near Old Babbitt, supports the common belief that the adjacent Animikie group sediments were nonconformably deposited upon the slightly irregular eroded surface of a granite. Granitic pebbles and boulders within the Pokegama and Biwabik formations strongly support this interpretation, though it would be difficult to prove that these granitic fragments did not originate in the vast granitic and metamorphic terrain immediately to the north because of the additional presence of boulders of a variety of metamorphic rocks. Core specimens of the granite-Pokegama formation contact also suggest a depositional contact. Some thin sections of this contact, however, show peculiar mineralogical variations near the contact. For example, in core from hole 32 the Pokegama formation consists of a fine-grained mosaic of approximately 60 per cent quartz, 25 per cent orthoclase, and 15 per cent biotite which is locally aggregated into lamellae-like concentrations. It also contains some irregular, slightly poikilitic grains of a dark green hornblende that possesses the same optical properties as the hornblende occurring in the granite. Although none of the silicate grains appear to be clastic, the decussate fabric of the grains suggests metamorphic recrystallization in the absence of stress. The nature of the granite in this core is essentially the same as described above, but the distribution of the silicates away from the contact with the Pokegama formation is significant. Within about 2 cm along the contact with the Pokegama formation the granite contains mafic clusters of hornblende, sphene, apatite, some ilmenite(?), and only insignificant amounts of biotite. With greater depth from this zone, hornblende is present only as small remnants within locally abundant amounts of brown and green biotite that contain numerous pleochroic haloes. Some sphene persists in this zone, but in most cases diamond-shaped skeletal networks of fine-grained ilmenite(?) now occur in its place. In other places the overlying sediments adjacent to the granite locally contain minerals, such as andradite, plagioclase, orthoclase, hypersthene, hornblende, apatite, and biotite, that are perhaps somewhat more typical of contact metamorphic mineral assemblages. The geometric relationships seen at some outcrops of the granite and the Pokegama seem to suggest an intrusive contact also.

In summary, it is probable that the Animikie group sediments were deposited on an irregularly eroded granitic mass but that a subsequent metamorphic event locally affected the textures and mineral assemblages of this granite and the immediately adjacent sediments. There is no evidence apparent to the writers to indicate whether this metamorphism was a result or not of the intensive igneous activity accompanying the emplacement of the Duluth gabbro.

POKEGAMA FORMATION

Except near Old Babbitt, there are few outcrops of the Pokegama formation that are readily available for examination. The upper part of the Pokegama formation was entered in holes 21 and 27E and was cut completely in holes 17, 32, and 34, in which the thicknesses of the formation are found to be 13, 30, and 48 feet, respectively. The formation appears to thin appreciably northeastward along strike and up-dip and is apparently missing beyond the eastern end of the Mitchell pit. The Pokegama formation is also apparently missing in core from hole 26; however, it is possible that the bottom foot, a chloritic quartzose conglomeratic zone within the iron formation, might correlate with the Pokegama. The Pokegama is generally a whitish- to greenish-gray quartzite. The basal beds of the Biwabik formation are also quartzose but they usually have a pronounced pebbly or conglomeratic fabric and commonly contain some magnetite and silicates that are typical of the iron formation.

Not a simple, homogeneous quartzite, the Pokegama formation actually consists of a variety of quartzite rock types. The most common is a fine- to medium-grained, massive, greenish-gray quartzite. Many thin sections of this variety show relic "sandy" fabrics that illustrate the original clastic nature of the Pokegama formation. In these rocks the rounded "sand grains" are now quartz, though some might have been clastic chert grains prior to recrystallization. Almost all the grains are clear and many possess clear quartz overgrowths. Most of the interstitial cement is not clear quartz and generally consists of a variety of minerals, particularly green and brown biotite. In some places the rock contains up to 10 per cent of apatite, andradite, actinolite, cummingtonite, and locally a trace of blue tourmaline, as interstitial minerals. In most places the biotites and amphiboles contain many pleochroic haloes around minute, enclosed zircon grains, and both varieties of biotite have partly altered to chlorite, numerous grains of which still contain subdued pleochroic haloes inherited from the biotites. Although most of the above silicates are essentially interstitial, many of the clastic quartz grains have been partly to almost completely replaced by these silicates. Some thin sections of this most common rock type show few relic sedimentary fabrics because of rather complete recrystallization, which has produced a granoblastic texture. In these cases the rock consists of xenoblastic quartz with differing amounts of interstitial brown and green biotite and secondary chlorite.

In many places the formation contains thin (down to $\frac{1}{16}$ inch) to thick (up to 6 feet) zones of fine-grained, laminated, dark green quartzite. These rocks contain abundant green and some brown biotite within lamellar segregations that probably reflect relic layering. Both biotites generally contain pleochroic haloes. The fabric of the biotite in these lamellae is distinctly decussate rather than lepidoblastic. Some fine-grained pyrrhotite also occurs in trace amounts with the biotite.

Creamy white quartzite layers, ranging from 1 inch to 30 feet thick, appear locally in core from holes 32 and 34. In thin section this variety of quartzite consists of a granoblastic mosaic mostly of quartz and, locally, up to about 40 per cent of silicic feldspar (orthoclase?). Notably silicate-poor as a rule, the quartzite locally contains many minute interstitial grains of zircon, chlorite (or possibly green biotite), muscovite, and, rarely, a green hornblende. In several places this variety of quartzite contains numerous thin, biotite-rich laminated zones that are essentially identical to those described directly above.

Pink andradite occurs to a minor extent as small porphyroblasts and thin veinlets locally within the silicate-rich parts of the greenish-gray quartzite near the top of the formation in core from holes 17 and 21 and just above the granite contact in hole 17. The greatest variety of silicates within the quartzite occurs near the top of the unit in core from hole 17. Andradite and highly pleochroic hypersthene occur as distinct remnants in a granoblastic mosaic of quartz, brown biotite, cummingtonite, and a few grains of plagioclase (oligoclase?). The latter group of minerals appears to have formed essentially contemporaneously. Both biotite and cummingtonite locally contain minute radioactive minerals (zircon?) that have produced pleochroic haloes. Incipient alteration of garnet and biotite to pale greenish chlorite has taken place. Elsewhere in core from this hole, medium- to coarse-grained hornblende within the quartzite has been almost completely altered to fine-grained nontronite.

The arkosic and shaly varieties of quartzite that occur locally to a minor extent within the Pokegama formation on the Main Mesabi range were not found on the Eastern Mesabi range. The creamy-white feldspathic quartzites and the greenish-gray, laminated, biotitic quartzites found in the eastern district, however, might be their metamorphic equivalents.

Contact between the Biwabik and Pokegama Formations

The contact between the iron formation and the quartzite cannot be located exactly in the highly reconstituted core of hole 17. However, the

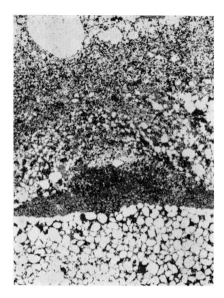

FIGURE 6. — Contact of the Pokegama quartzite and the Biwabik iron-formation. The basal layers of the Biwabik iron-formation contain abundant clastic "sand and pebble" grains of probable extraformational origin. The interstitial minerals consist of fine-grained quartz, green and brown biotite (and secondary chlorite), andradite, apatite, and many small zircon grains. The dark, thin, layer-like mass at the bottom of the formation consists of a fine-grained granoblastic mosaic of andradite, apatite, and brown biotite with numerous pleochroic haloes around minute zircon grains. The underlying Pokegama formation is essentially quartzite consisting of clastic quartz grains with interstitial fine-grained quartz, andradite, apatite, and green and brown biotite, with secondary chlorite. (21-475½; 4X)*

FIGURE 7. — Contact of the Pokegama quartzite and Biwabik iron-formation. The "sand and pebble" clastic grains of the Biwabik iron-formation are surrounded by interstitial fine-grained actinolite, green and some brown biotite, quartz, blue tourmaline, and some cummingtonite. The clastic pebble in the upper middle of the picture has been completely replaced by these minerals. The Pokegama formation is shaly at this locality. The abundant fine-grained clastic quartz grains are also associated with interstitial fine-grained actinolite, green and brown biotite, blue tourmaline, and some cummingtonite. Both biotite and tourmaline form in thin lamellae-like aggregates which probably reflect initial bedding structures. (32-449, 4X)

contact is generally sharp and easy to find in cores from other holes. Three different but common varieties of this formational contact are shown in Figures 6 to 8. Each case is considered to be a conformable sedimentary

* (21-475½; 4X) indicates that the illustrated thin section is of core from hole 21 at a depth of 475½ feet. 4X indicates the linear magnification of the photomicrograph. Unless otherwise noted, all photomicrographs are of thin section photographed in ordinary light.

contact, although each displays slightly different relic textures and thereby warrants separate illustration.

The relic "sandy" fabric of the greenish-gray quartzite in contact with the relic "pebbly" fabric of the basal beds of the iron formation, found in core from hole 21, is shown in Figure 6. The outlines of the clastic grains of both formations are accentuated by the presence of fine-grained green and brown biotite, secondary chlorite, andradite, and apatite. Some of the biotite is associated with minute specks of a cream-colored opaque substance, probably leucoxene, and both types of biotite contain appreciable numbers of pleochroic haloes which are still apparent in the secondary chlorite.

In hole 32 the pebbly basal layer of the iron formation lies upon one of the laminated silicate-rich quartzite zones of the Pokegama formation as illustrated in Figure 7. In this photograph the clastic grain outlines and lamellae are emphasized again by the presence of abundant silicates. The material interstitial to the relic "sand grains" and "pebbles" is essentially the same in both units and consists mainly of fine-grained, pale green-yellow actinolite, green biotite, pale blue tourmaline, some brown biotite, and minor cummingtonite. The biotites and actinolite of both formations contain pleochroic haloes. The green and brown biotite, cummingtonite, actinolite, and tourmaline are intimately associated and appear to be contemporaneous, although in many places some textures indicate that cummingtonite might possibly be later. Relic quartzose pebbles and granules are obviously abundant in the iron formation of the upper part of the picture and locally many of them have been partly or almost wholly replaced by biotite, actinolite, cummingtonite, and tourmaline, and in a few instances these silicates are enclosed within the quartz grains. The dark pebble just above the center of the picture now consists in the main of these silicates.

The Pokegama quartzite in hole 27E is conglomeratic, and the base of the iron formation consists of distinctly layered chert beds as shown in Figure 8. There is a remote possibility that this conglomeratic quartzite is in reality the pebbly basal bed of the iron formation because the drilling terminated in the unit. It is apparently identical, however, to the Pokegama obtained from other nearby drill holes and has also been identified by Erie Mining Company geologists as Pokegama quartzite. Many of the clastic grains of the quartzite are almost entirely replaced by fine-grained chlorite, but others are only partly replaced. In several instances the chlorite also exists in the adjacent matrix and in still others it exists only interstitially to the present quartz grain boundaries. In some chlorite-rich grains minute quantities of interstitial carbonate are also present. Many chlorite clusters contain darker pleochroic spots that resemble, and are presumably inherited from, those noted in the biotite from the holes farther east. Minor amounts of chlorite are also present in quartzose granules and along the stylolites within the bedded chert layers of the iron formation. A single grain of brown biotite, with some pleo-

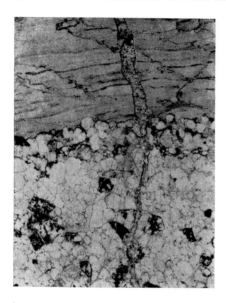

FIGURE 8. — Contact of the Pokegama quartzite and Biwabik iron-formation. In this specimen the basal strata of the Biwabik formation consist mainly of bedded chert with only minor amounts of fine clastic quartz grains. Traces of chlorite are present in the clastic grains and in a few thin discontinuous veinlets. The underlying Pokegama formation consists largely of clastic quartz grains with clear secondary overgrowths. The interstitial material is very fine-grained quartz and chlorite, and rarely remnants of brown biotite. Some of the clastic quartz grains have been largely replaced by chlorite. Both formations are cut by a veinlet of fine-grained quartz, chlorite, and carbonate. (27E-561; 4X)

chroic haloes and completely enclosed within a quartz grain, is only partly altered to pale green chlorite. The thin veinlets that cut the specimen consist mainly of quartz, pyrite, chlorite, and a carbonate.

BIWABIK IRON-FORMATION
GENERAL FEATURES

The division of the Biwabik iron-formation into the so-called Upper Slaty, Upper Cherty, Lower Slaty (including the Intermediate Slate), and Lower Cherty (Wolff, 1917) members has been retained, although these members have been subdivided further into many submembers. In the following discussion these submembers are commonly referred to as "units." The columnar section of the Biwabik formation is presented in Figure 5. The correlation of the submembers to the well-known iron formation members, as well as to the units proposed by Grout and Broderick (1919), is also presented in this column. A longitudinal diagram showing the distribution of the members and submembers of the Biwabik formation on the Eastern Mesabi is presented with the map accompanying this report.

The lithologic classification of taconite was discussed in Chapter 1. There it was pointed out that the taconites in this district consist mainly of two distinctly different kinds of strata, e.g., bedded taconite-strata which are largely responsible for the bedded or thinly layered aspects of the taconite and massive taconite-strata which are massive and lacking in internal stratification. In the descriptive text below, it was found convenient to use these taconite-strata terms in describing the structures and mineral assemblages occurring in each of the stratigraphic units. Certain other terms are also used in the following manner. The term "zone" indicates only "a thickness of rock," which generally displays a particular feature or features. The terms "magnetite-rich" and "silicate-rich" as used in the descriptive text below are merely comparative terms. For example, a particular thin zone of rock containing many magnetite-bearing lamellae or magnetite-bearing pebbles may be described as a "magnetite-rich" portion of a given specimen, which as a whole may be highly quartzose. Economically speaking, only some of the submembers, mainly from the Upper Cherty member, are sufficiently rich in magnetite to be amenable to beneficiation with present methods. If a mineral is described as being locally abundant, it is implied that the mineral locally constitutes approximately 30 per cent or more of the particular core or rock specimen being discussed. In a similar manner, other adjectives are used to indicate the visual estimates of the amount of mineral content: they are moderate (10 to 30%), minor (2 to 10%) and trace (less than 2%) amounts. In describing the grain size (diameter) of the minerals of the iron formation, the following size ranges are implied: very fine-grained (less than .05 mm), fine-grained (.05 to .25 mm), medium-grained (.25 to 1 mm), coarse-grained (1 to 10 mm) and very coarse-grained (greater than 10 mm). These coarser-fabric rocks are not foliated nor lineated and can be described as granoblastic hornfelses, or preferably as granofelses as defined by Goldsmith (1959, p. 109).

The rocks have, in general, a "granular texture" because of the coarsened fabric produced by recrystallization. Due to the presence of small "granule structures," many of the quartzose layers throughout the formation may have in addition a "granule texture." These granule structures are generally well-rounded subspheroidal quartzose masses that range from the size of coarse-grained sand to very small pebbles. They are typically oval-shaped when viewed in sections cut perpendicular to the bedding planes of the rock (Fig. 29). Most range from 0.5 to 2 mm in diameter, but some small pebbles may be of identical origin. On the Eastern Mesabi most of the granules are chert-like quartzose masses that contain differing amounts, from zero to near 100 per cent, of micron-sized magnetite dust (Fig. 40, A). Many relic granule structures are also preserved in places where the quartz and magnetite have been slightly coarsened by recrystallization. In a few instances, hematite (martite?) is similarly distributed within granules. Minute amounts of extremely finely divided clay-like material, resembling allophane, are apparent in many of the highly quartzose granules. Many relic granule structures are still evident in outline even where clusters of metamorphic silicates have extensively replaced them. Granule-shaped clusters of minute inclusions within some silicates have also been interpreted to be relics of these structures. Relic granule structures on the Main Mesabi range are best described as masses of fine-grained silicates, especially greenalite, minnesotaite, and stilpnomelane; granule structures composed of fine-grained chert and dusty magnetite are rare. The writers believe the granule structures of the entire

Mesabi range were probably precipitated flocs of chemical origin, somewhat resembling silica gel, that were subsequently congealed, rolled, and well-sorted on the ocean floor to produce the obvious clastic fabrics that the granules and their fragments produce. A fuller treatment of the relic sedimentary structures, especially granules, of the iron formation will be given in a later paper.

For the purpose of discussion of lithologic features observed in drill core specimens, all holes east of Rangeline 13 West and including hole 21 are, in general, collectively called the eastern holes. The remaining holes, excluding hole 27E near Mesaba, are generally referred to as the western holes. Exceptions to this grouping will be noted where appropriate, as in the Lower Cherty member discussed below.

Most of the available drill core was obtained from the upper members of the iron formation; in fact, cores from the entire section were not available during the first stages of the study. Consequently, for convenience, the submembers are lettered in succession as encountered in drilling, the uppermost unit being submember A and the basal unit submember V.

Lower Cherty Member

The Lower Cherty member has been tentatively divided into submembers R, S, T, U, and V. This member consists essentially of a thin wedge of sediments that apparently pinches out to the northeast. There are many bedding features that occur in cores from some holes but not in those from other holes, as might be expected in such a stratigraphic unit. The problem of description and correlation is made more complex by the effects of metamorphism on the sediments and the paucity of available specimens. Core from the member was obtained only from holes 17, 21, 26, 32, 34, and 27E; the thickness of the member at these places is 31, 49, 32, 45, 50, and 136 feet, respectively. Because the holes are so widely separated (about 12 miles from 17 to 27E) within a member that varies greatly in thickness, the proposed submembers of the Lower Cherty are not as well defined nor as easily correlated as other submembers of the formation. Although the other submembers of the formation are also heterogeneous in detail, they are lithologically rather uniform in their areal extent and therefore could be relatively easily recognized and correlated from hole to hole, in spite of variations in silicates assemblages present.

The upper contact of the member with the Lower Slaty member is easily identified by the normally abrupt change to the silicate-rich, magnetite-poor, dark unit Q (The "Intermediate Slate") from the more quartzose, layered, magnetite-bearing taconite of the Lower Cherty member. The lower contact of the member is normally placed below the lowest magnetite-bearing taconite stratum, which is commonly marked by a quartz pebble conglomeratic zone lying above the Pokegama formation, as described above.

For the purposes of discussion of the Lower Cherty member only, hole

27E alone is called the western hole and the remaining five holes on the Reserve Mining Company properties are called the eastern holes. There are parts of this member that are somewhat similar in cores from hole to hole, and although they might be only approximately equivalent and lacking in uniformity, it is helpful to discuss these tentative submembers. The magnetite and quartz of these units are mainly fine-grained throughout the core from each hole.

Submember V. This quartzose basal unit is generally magnetite-poor, and except for the presence of granule structures and silicates typical of the iron formation, it locally resembles the underlying Pokegama formation. Unit V was not recognized in core from hole 17 but is $3\frac{1}{2}$, 11, $1\frac{1}{2}$, 12, and 33 feet thick in holes 21, 26, 32, 34, and 27E, respectively.

The unit consists dominantly of massive quartz taconite with minor amounts of magnetite-bearing lamellae, thin layers, and granules. In core from the eastern holes minor to locally abundant amounts of fine- to coarse-grained hedenbergite, and rarely traces of andradite, occur in the quartzose massive taconite-strata. Minor amounts of cummingtonite also occur in these quartzose layers, and have generally replaced hedenbergite to some degree. The less abundant bedded taconite-strata of the eastern holes consist largely of fine-grained magnetite, but in hole 27E the magnetite-bearing layers are commonly associated with stilpnomelane in minor amounts. In this westernmost hole the granule structures of the quartzose layers are commonly accentuated by the development of fine-grained minnesotaite and green stilpnomelane within them. In a few places porphyroblasts of carbonate give the core a mottled appearance. In one place the core from hole 27E contains thin dilated veinlets of brown stilpnomelane. Here, fine-grained lamellae of minnesotaite and carbonate and layers of minnesotaite and quartz have been largely replaced by brown stilpnomelane adjacent to the veinlet.

The bottom eight inches or so of the cores of this unit are generally conglomeratic and contain numerous quartzose pebbles that are probably extraformational in origin. An exception to this generality is found in core from hole 34 where the conglomeratic zone is 8 feet thick; the presence of granule structures in this thick conglomeratic zone was the basis for not classifying it as Pokegama formation. The conglomeratic zone in the cores from all the holes probably corresponds with the hematitic "red basal taconite" of the Main Mesabi range, although no hematite was observed in the Eastern Mesabi district.

The grain fabrics of the basal part of this unit are illustrated in the discussion of the underlying Pokegama formation (see pp. 20, 22). The mineralogy of the very basal beds is the same as that of the remainder of the unit; but in a few extensively recrystallized specimens, such minerals as andradite, apatite, blue tourmaline, actinolite, zircon, green and brown biotite (and secondary chlorite) occur in minor amounts.

Submember U. This unit locally contains appreciable amounts of magnetite, especially in granule structures and groups of lamellae, which occur within silicate-bearing quartzose layers. In holes 17, 21, 32, and 27E, unit U is 4, 9½, 15½, and 57 feet thick, respectively. Units U and T could not be differentiated in the cores from holes 26 and 34; in these places the combined thicknesses of these units are 9½ and 22 feet, respectively.

In the eastern holes magnetite-rich granules are commonly packed closely together and are locally somewhat segregated into layered zones that range up to 4 inches in thickness. Local concentrations of magnetite lamellae also occur in thin laminated zones up to ¾ inch thick. Except in a few places, silicates have not developed abundantly within the magnetite-rich parts of the rock. In the western hole, 27E fine-grained stilpnomelane is locally abundant within and adjacent to the magnetite-rich laminated beds along with minor amounts of minnesotaite and carbonate. These magnetic layers are particularly stilpnomelane-rich where the rock has been cut by a number of thin stilpnomelane-pyrite-calcite-quartz veinlets, some of which are cut in turn by very thin veinlets of stilpnomelane. Moderate amounts of pyrite and minor amounts of chalcopyrite occur immediately adjacent to the sulfide-bearing veinlets in a few places. In general, the abundance of stilpnomelane in the wallrock decreases away from the veinlets.

The interbedded massive taconite-strata generally range from ¼ to 12 inches but are mostly 1 to 5 inches thick. These layers are generally quartzose throughout the district but locally contain substantial amounts of silicates. In some of the eastern holes minor amounts of fine-grained fayalite occur as remnants within the locally abundant fine- to coarse-grained hedenbergite, which occurs in most holes. In several places the pyroxene is accompanied by trace to minor amounts of andradite, calcite, and pyrrhotite. Starting with core from hole 17 and proceeding westward, both fayalite and hedenbergite have been replaced by increasing amounts of fine-grained cummingtonite, which is locally accompanied by minor amounts of hornblende+cummingtonite and actinolite, particularly where replacing hedenbergite. Some thin sections also show fairly well-preserved, dusty, magnetite-bearing granules that have been incipiently replaced by small bundles of acicular cummingtonite and trace amounts of hornblende+cummingtonite. Cores from holes 21 and 32 contain numerous quartz grains, as seen in Figure 9, that have been interpreted to be relic clastic grains of quartz or possibly of recrystallized chert. These clastic-like grains are surrounded and emphasized by the dark green matrix, which consists in part of hedenbergite along with some oligoclase, hornblende, calcite, apatite, and fayalite — all of which have been substantially replaced by acicular hornblende+cummingtonite and cummingtonite containing many pleochroic haloes. These minor clastic zones show fabrics nearly identical to those found in parts of the Pokegama formation. In hole 27E to the west, abundant amounts of fine-grained minnesotaite as well as some carbonate and stilpnomelane locally accentuate the granule structures of the quartzose layers.

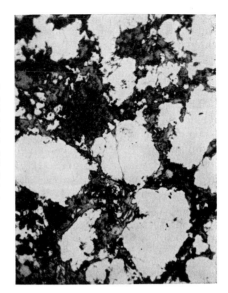

FIGURE 9. — Clastic fabrics in unit U. The coarse grains of quartz are relic of clastic, coarse-sand-size fragments which might have been initially chert and, or, quartz. The interstitial matrix to the grains consists of medium- to fine-grained hedenbergite, oligoclase, hornblende, calcite, apatite, and traces of fayalite. Many of the above silicates and some of the "sand" grains have been partly replaced by fine-grained hornblende+cummingtonite and cummingtonite. (21-463; 20X)

Submember T. This unit contains moderate to locally abundant amounts of magnetite, especially in the form of granules, within essentially quartzose taconite. Cores indicate that unit T is 2, 7, and 11½ feet thick, respectively, in holes 21, 32, and 27E. Undifferentiated units S + T are 18 feet thick in hole 17, and as mentioned above, units U and T could not be distinguished separately in cores from holes 26 and 34. In a general way, the distribution of magnetite consists of numerous irregular zones of densely-packed magnetite-rich granules near the top, but changes to more clearly defined thin layers and lamellae of magnetite nearer the bottom. In many respects the distribution of the magnetite is somewhat similar to that in unit J of the Upper Cherty member. Except for a few hisingerite veinlets in core from hole 32, silicates are commonly sparse among the magnetite-rich parts of the unit.

The massive taconite-strata from this unit are generally quartzose everywhere, with only subordinate amounts of silicates having developed. Cores from holes 17 and 21 contain minor amounts of fayalite remnants in the quartzose layers. Nearly all the eastern holes yielded cores containing moderate amounts of fine-grained hedenbergite, which has been slightly replaced by fine-grained cummingtonite in most places. One core from hole 21 is cut by a thin quartzose veinlet containing some pyrrhotite and hedenbergite. To the west in core from hole 27E, fine-grained minnesotaite is generally abundant within the granule structures and in most places within the surrounding quartzose matrix to these structures. Small porphyroblasts of carbonate give the quartzose layers in the core a mottled appearance.

Submember S. This unit contains moderate amounts of magnetite, mainly as thin layers, within essentially quartzose taconite. It resembles the overlying unit R except that quartz rather than silicates dominate the nonmagnetic parts of the rock. The thickness of this unit in holes 17, 21, 26, 32, and 27E are, in order, 13, 4½, 9, 8, and 22 feet. Units R and S could not be separated in core from hole 34; here their combined thickness is 16 feet.

The bedded taconite-strata of this unit consist mainly of thin magnetite-rich layers that commonly range from ¼ to ½ inch thick, and to a lesser extent by laminated zones of similar thickness that are also rich in magnetite. In a few places magnetite-rich granules are closely packed into irregular layer-like zones. In general, only minor amounts of silicates are found among the magnetite-rich parts of the rock.

The interlayered massive taconite-strata, most commonly ¼ to 2 inches thick, are generally quartzose and contain abundant granule structures in most of the cores. In core from hole 17, these structures have been obliterated during recrystallization, assuming they were once present. Algal structures and their fragments, defined by fine-grained magnetite in quartz, were found near the top of the unit only in core from holes 21, 26, and 32. Minor amounts of fine-grained fayalite occur, mainly as remnants within hedenbergite and cummingtonite, in core from holes 17 and 21. Minor to moderate amounts of fine- to coarse-grained hedenbergite occur in the quartzose layers of all the eastern cores. In most places, the olivine and pyroxene have been partly replaced by fine-grained cummingtonite and some hornblende+cummingtonite. In numerous other places acicular bundles of fine-grained cummingtonite have incipiently replaced granule structures consisting of fine-grained magnetite in quartz.

Several thin hisingerite veinlets cut the core from holes 17 and 21. Where these cores contain hedenbergite, the pyroxene is slightly darkened and altered to brownish nontronite next to the vein. In other parts of the same cores fayalite is commonly altered to brown and green hisingerite adjacent to the veinlets.

A quartzose taconite specimen from hole 21 is unusual because it contains the only occurrence of barite recognized in the district. It occurs in a granoblastic mosaic along with quartz, calcite, stilpnomelane, and feldspar(?). The calcite and especially the barite have been partly replaced along their cleavages by fine-grained acicular stilpnomelane.

In the one western hole, 27E, distinct granule outlines preserved by opaque cloudy material (resembling allophane?) are visible within the large metacrysts of calcite and ankerite — ranging up to 8 mm in diameter — which give the core a mottled appearance. Minor amounts of fine-grained, pleochroic yellow-green minnesotaite rim the carbonates in many places. Elsewhere in these cores, minor amounts of acicular fine-grained colorless minnesotaite as well as green and brown stilpnomelane have incipiently developed within clouded cherty granules.

Submember R. This uppermost unit of the Lower Cherty member contains the last magnetite-rich layers and layered zones to form before the deposition of the overlying magnetite-poor "Intermediate Slate" (unit Q). It is 8, 12½, 18½, 13, and 12½ feet thick respectively in holes 17, 21, 26, 32, and 27E. The combined thickness of units R and S in core from hole 34, where these units could not be differentiated, is 16 feet.

The magnetite-rich portions of the unit are relatively few in number and generally consist of thinly layered and laminated zones that average about ¼ to ½ inch thick, although some reach 4 inches in thickness. Core from hole 27E contains, in addition, numerous layer-like zones composed of closely-packed, magnetite-rich granules. Throughout the district the silicates that occur in the interlayered massive taconite-strata of a given core are locally abundant in the magnetite-bearing parts of the core.

The massive taconite-strata of this unit are normally 1 to 8 inches thick and generally consist of locally silicate-rich quartzose layers. In most of the eastern holes moderate to locally abundant amounts of fine-grained fayalite occur as remnants within lesser amounts of fine- to coarse-grained hedenbergite. In core from hole 17, minor amounts of ferrohypersthene are associated with the fayalite. All these silicates have been replaced to some extent by fine-grained cummingtonite, especially in core from hole 32. In one specimen, however, medium-grained hedenbergite containing relic granule structures consisting of thin rims of fine-grained magnetite has been replaced by fine-grained actinolite. In this specimen and in other cores from most of the eastern holes minor amounts of disseminated pyrrhotite occur within and interstitial to hedenbergite, and in core from hole 32 it is locally abundant and associated with minor amounts of chalcopyrite.

An unusual mineralogical association occurs in core from hole 21, illustrated in Figure 10. Here porphyroblasts of almandine (2 to 5 mm in diameter) occur within and have been slightly replaced by a matrix of fine-grained green and brown biotite and fine- to medium-grained cummingtonite, both of which contain many pleochroic haloes around minute zircon grains. In core from hole 32, minor amounts of andradite and calcite occur interstitially to hedenbergite.

In the core from the western hole, 27E, the granule structures within the quartzose layers are mostly composed of fine-grained minnesotaite and to a minor extent of stilpnomelane and siderite, but where these minerals are extensively developed, the granule structures are not particularly well preserved. There are generally a few places, however, where these minerals have only incipiently developed and the relic granules are well preserved and consist largely of considerably finer-grained quartz that generally contains cloudy, white, opaque material.

In hole 27E a transition zone about 10 feet thick occurs between the Lower Cherty and Lower Slaty members. This zone consists of interbedded rock-types somewhat similar to those of units Q and R. One type consists largely of thinly layered and laminated zones composed of differing amounts of stilpnomelane, minnesotaite, carbonate, magnetite, and quartz. The other type normally consists of silicate-bearing quartzose

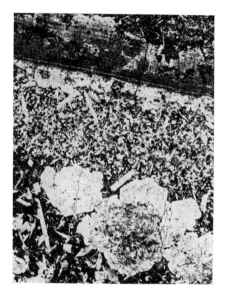

FIGURE 10. — Almandine porphyroblasts of unit R. Near the top of the photograph relic laminations are preserved (by graphite?) within a layer-like zone of almandite. Above this, numerous small remnants of fayalite are almost completely replaced by fine-grained cummingtonite and green and brown biotite. In the lower portions of the picture several coarse subhedral porphyroblasts of slightly birefringent almandite are incipiently replaced by fine- to medium-grained cummingtonite. The dark interstitial matrix consists largely of brown biotite, silicic plagioclase, and quartz. (21-445; 5X)

layers containing numerous granule structures that are generally quartzose and silicate-poor. Although rather sharp contacts between the two rock types are common, Figure 11 shows one of the many instances where considerable numbers of granule structures adjacent to the layered and laminated zones have been partly to entirely replaced by stilpnomelane, minnesotaite, and carbonate. The degree of metasomatic replacement decreases uniformly away from the laminated zones, as can be seen in the

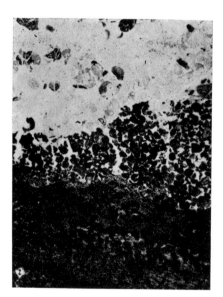

FIGURE 11. — Granule structures replaced by stilpnomelane. In this specimen, solutions moving along a group of magnetite lamellae (bottom, black) slightly recrystallized much of the magnetite and deposited new interstitial mineral assemblages consisting largely of stilpnomelane, carbonate, and traces of minnesotaite. Locally the solutions permeated the adjacent granule-bearing layers, and completely replaced the nearby granules mainly by stilpnomelane (center, dark gray), carbonate, and minor amounts of minnesotaite. Partly replaced granules line the replacement front. Quartzose granules (top), containing only an incipient development of stilpnomelane (gray) and carbonate, are found in the rest of the adjacent layers. (27E-417½; 3X)

photograph. Here fluids moving along the channelways within the laminated beds permeated outward into the quartzose layers and there were able to reconstitute the existing material of the granules into the new mineral assemblages.

Lower Slaty Member

The member has been divided into two submembers, units P and Q, the latter of which corresponds to the so-called Intermediate Slate (or the "Paint Rock" where altered) of the Main Mesabi range. The Lower Slaty member generally contains less total iron, and iron as magnetite, than the other members of the iron formation. It is typically silicate-rich and poorly bedded, or laminated, throughout the district. The upper part of the Lower Slaty member, unit P, normally contains silicate assemblages similar to those near the base of the Upper Cherty member. Consequently, the upper boundary of the member is generally placed at the appearance of abundant magnetite-rich beds or granule structures and the well defined, bedded, quartzose layers presumed to belong within the Upper Cherty member. This contact is generally less easily recognized and is determined more subjectively on the Main Mesabi range, particularly where the formation is altered and oxidized.

The Lower Slaty member is normally uniform in thickness, ranging from 78 feet near Mesaba to an average of about 86 feet in the Eastern Mesabi district, but it is 117 feet thick in core from hole 17. Most of the exploration drill holes in the Eastern Mesabi district terminate within the upper part of submember P.

Submember Q. This magnetite-poor unit is characteristically much darker than the adjacent units and is easily recognized on that basis. Cores from holes 17, 21, 26, 32, 34, and 27E show that the submember is $19\frac{1}{2}$, $32\frac{1}{2}$, 27, $23\frac{1}{2}$, 22, and 20 feet thick, respectively, at these localities.

With the exception of holes 17 and 21, where the core is lighter in color and generally contains coarse-grained silicates, submember Q can be described as a black to dark gray, fine-grained, graphitic silicate- and quartz-bearing slate-like rock, although it lacks a true slaty cleavage. It generally appears to be massive but is frequently indistinctly layered and laminated, and the bedding surfaces of these layers have a graphitic sheen. Particularly to the west, there is often an inherent bedding-plane weakness that causes the rock to break into slabby or tabular fragments. In core from hole 26, which is up-dip from the other holes, the unit is distinctly laminated and locally contains numerous thin chert lamellae and lenticular layers up to $\frac{1}{8}$ inch thick. Cores from the four holes west of hole 21 also commonly contained locally prominent conglomeratic zones consisting of rounded as well as tabular or slab-like chert fragments, around which minor amounts of fine-grained pyrite have formed (Fig. 12). Similar appearing conglomeratic pebbles in core from hole 32 have been largely replaced by fine-grained cummingtonite and pyrrhotite. These conglomeratic zones are undoubtedly of intraformational origin. This

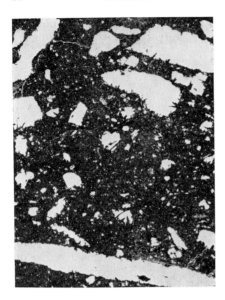

FIGURE 12. — Intraformational chert conglomerate of unit Q. Subangular to tabular fragments or pebbles of fine-grained chert-like quartz in a dark, very fine-grained matrix of quartz, graphite(?), and pyrite form a distinctive conglomeratic zone in unit Q. Locally small subhedral prisms of pyrite and acicular stilpnomelane replace the chert fragments along their edges. (27E-410; 3X)

variety of intraformational conglomerate is common in the region of the Main Mesabi range west of Mesaba, where the chert-like fragments commonly contain numerous granule structures.

The dark carbonaceous material of this unit cannot be identified with the hand lens nor in thin section. The presence of extremely fine-grained quartz has made it virtually impossible to detect the presence of graphite by x-ray diffraction.

Several features characterize unit Q in cores from holes 17 and 21, the most important being the presence of coarse-grained silicates and thin dark lamellae rich in graphitic(?) material. These dark lamellae are not magnetite nor sulfides, and in thin section can only be described as extremely fine-grained, disseminated, opaque, carbonaceous material, presumably graphite.

The dominant silicate in the cores from these two eastern holes is medium- to coarse-grained ferrohypersthene, ranging up to 2 cm in diameter in core from hole 21. It commonly occurs as very coarse-grained, irregular, poikiloblastic metacrysts, or massive zones of such grains, that have developed within a poorly laminated granoblastic mosaic of fine-grained fayalite, quartz, graphite, and traces of pyrrhotite. The massive layer-like zones of ferrohypersthene are commonly rather abruptly bounded above and below by the darker graphitic layers, so that the rock consists essentially of an irregularly alternating layered sequence of darker fine-grained (fayalite)-quartz-graphite-rich layers and of lighter coarse-grained layered zones rich in ferrohypersthene (Fig. 13). These alternating dark and light layers are generally between $\frac{1}{8}$ and $\frac{1}{4}$ inches thick and rarely reach one inch in thickness. All of the light and dark layers doubtless represent primary bedding although perhaps much of the layer-

FIGURE 13. — Graphite-ferrohypersthene fabrics in unit Q. Metacrysts of coarse-grained ferrohypersthene (light gray) occur within the carbonaceous, quartzose "Intermediate Slate." Here the carbonaceous material, now graphite (black) could not be incorporated into the crystal and was pushed aside by, and thus concentrated along one side of, the growing crystals. Minor amounts of fine-grained fayalite and traces of pyrrhotite occur in the light-colored granoblastic quartzose zones of the rock. (21-435; 9X)

ing has been modified during the development of the metamorphic silicates. A few specimens show almost entire reconstitution of the rock into ferrohypersthene and fayalite "bands" in which relic layering of opaque material is apparent in thin section but is rather subdued in hand specimen. In these specimens fayalite occurs as remnants within the pyroxene zones or as separate layered zones between them. In most of the cores containing these silicates both the fayalite and ferrohypersthene are replaced to a minor extent by fine-grained cummingtonite and hornblende+cummingtonite. Textural relations shown by some poikilitic cummingtonite grains, however, might be interpreted to mean that the development of some of the cummingtonite is essentially contemporaneous with the earlier anhydrous silicates. At the other extreme, a specimen from hole 21 shows only an incipient development of fayalite anhedra within very fine-grained graphitic quartzose material, although some poikilitic ferrohypersthene did develop at a later stage. Small subhedral crystals of a dark gray garnet, almandite, about $\frac{1}{2}$ to 1 mm diameter, locally occur within the graphite-quartz-rich matrix and within the fayalite and the ferrohypersthene, but the fabric of the grains is such that the relative age of the garnet is not clear. All the silicates in this rock, however, are partly replaced by fine-grained brown biotite and some cummingtonite, minor amounts of which also occur within the quartzose matrix.

Thin veinlets of hisingerite, with minor amounts of fine-grained magnetite, typically occur parallel or semiparallel to the relic layering in the cores from holes 17 and 21. Microscopic observations clearly show that the hisingerite veinlets are latest in relative age and that their orientation is perhaps related only indirectly to the bedding structure of the rock. Fayalite, and to lesser extent ferrohypersthene, are darkened and altered to hisingerite and nontronite next to the hisingerite veinlets, and hence in some cores supplement the "banded" appearance of the specimen.

Moderate to locally abundant amounts of pyrrhotite are common among the silicates, particularly ferrohypersthene, in many places.

Submember P. Unit P characteristically contains abundant amounts of many ferrous silicates but only minor amounts of magnetite. It is the most obviously metamorphosed and reconstituted submember of the iron formation in the district. Its minerals display many different metamorphic fabrics, and relic granule structures are still locally apparent in most places. Consequently, this unit was studied in detail in an attempt to determine the sequence of metamorphic mineral paragenesis for the area. The details of this study are discussed in the chapter on metamorphism (see pp. 104–109).

Cores from all holes except 1 to 3, 7, 12, 23, and 35 contain specimens of the upper strata of unit P. The entire thickness of the unit is represented, however, only in cores from holes 17, 21, 26, 32, 34, and 27E, where the unit is 97, 55½, 62, 62½, 57, and 58 feet thick, respectively.

Although numerous cores contain a few magnetite-rich lamellae, or even laminated zones up to 2 inches thick, the magnetite content of unit P rarely exceeds a few per cent. Some fine-grained magnetite also occurs in quartzose granule structures and a few small pebbles.

The massive taconite-strata, though largely quartzose in many occurrences, are generally characterized as silicate-rich in most places. The relic granule structures of these layers are largely quartzose or have been reconstituted into granule-shaped clusters of silicates during metamorphism. Most of unit P is poorly to massively bedded, but in numerous places fine-grained disseminated magnetite (and possibly some graphitic material) imparts a vaguely layered aspect to the silicates. As a rule, the silicate-rich parts of the cores are rather massive, though in several places small zones of silicate-rich material are distinctly thin-bedded.

The most abundant silicates found in this submember are fayalite, poikilitic cummingtonite and ferrohypersthene, and prismatic to acicular cummingtonite. The distribution of these silicates is essentially the same as that observed in the overlying unit O, as discussed below and in the chapter on metamorphism (see pp. 104–109). The distribution of the calcium-bearing ferrous silicates, hedenbergite and hornblende, is directly dependent upon the proximity to a metasomatic pegmatitic vein and, consequently, the occurrences of these silicates are not closely similar in units O and P. The lateral distribution of silicates within unit P is outlined below.

The rock constituting unit P in the cores from hole 18 and from holes to the east is a medium-grained, granoblastic, (magnetite)-quartz-fayalite granofels, i.e., a eulysite. A few poikilitic porphyroblasts of ferrohypersthene occur with the fayalite-bearing assemblages in the cores from holes 6, 9, 14, and 15. Moderate amounts of medium-grained hedenbergite have locally replaced parts of the eulysite rock in core from holes 8 and 18. The cores from several of these eastern holes also contain some poikilitic cummingtonite, but it cannot be determined if these grains developed with or after the formation of the minerals of the eulysite. In any event, all of the fayalite, poikilitic cummingtonite, poikilitic ferrohypersthene, and hedenbergite found in all these cores have been partly replaced locally by moderate amounts of fine-grained hornblende+cummingtonite and minor amounts of cummingtonite and hornblende. Minor amounts of fine-grained green and brown biotite and apatite accompany the hornblende and cummingtonite in the core from hole 18. All the fine-grained amphiboles and biotites commonly contain small pleochroic haloes around minute zircon grains. Thin magnetite-bearing hisingerite veinlets occur in the cores from holes 4 and 17, and fayalite adjacent to these veinlets is commonly altered to dark brown hisingerite or nontronite.

Specimens of unit P from the central group of holes (numbers 19 to 26) characteristically have a mottled appearance. The mottling is produced by roundish remnants of unreplaced fayalite-bearing mineral assemblages (similar to those occurring in the eulysites to the east) within a finer-grained groundmass consisting of prismatic cummingtonite, with or without xenoblastic quartz. A similar appearing, corroded remnant of a ferrohypersthene-bearing eulysite also occurs in core from hole 21. Minor amounts of hornblende+cummingtonite accompany, or have been slightly replaced by, the fine-grained cummingtonite in cores from holes 20, 21, and 25. Traces of plagioclase accompany cummingtonite in core from hole 24.

Very fine- to fine-grained prismatic to acicular cummingtonite, along with minor amounts of quartz, dominates the mineralogy of unit P in the cores from the holes remaining to the west. Small remnants of fayalite are found in trace amounts, however, in core from all of these holes except 31, 36, and 37. Traces of pyrrhotite occur with cummingtonite in core from hole 37. A few thin veinlets of hisingerite occur in core from hole 33.

Upper Cherty Member

The Upper Cherty member has been divided stratigraphically into eight submembers: H, I, J, K, L, M, N, and O. Most of these submembers contain pebbles of intraformational conglomerates as well as granules and thin beds that are rich in magnetite. In comparison with other members of the iron formation, the quartzose layers of this member are relatively thicker, distinctly bedded, and characteristically contain abundant relic granule structures in most places. Silicates are commonly found in many places in this member, but the three basal units contain locally abundant silicates in most places.

The Upper Cherty member was cut, at least in part, by all drill holes except numbers 1, 2, 3, and 7. From cores containing the entire Upper Cherty interval, the member was found to vary in thickness from about 120 to 160 feet, averaging about 140 feet, but it is about 246 feet thick near Mesaba.

This member generally contains the highest percentage of total iron, as well as the largest amount of iron occurring as magnetite. The bottom of the eastern end of the Peter Mitchell Pit of the Reserve Mining Company starts near the base of the Upper Cherty member and, proceeding westward, higher and higher horizons of the Upper Cherty member are exposed in the pit. The basal strata of the Upper Slaty member are also exposed and are currently being mined in the western parts of the Mitchell pit. The Erie Mining Company's pit number 2, east and slightly north of Mesaba, also exposes a larger part of the Upper Cherty section currently being mined. The silicate-rich basal units in the Babbitt area are generally avoided in mining, however, because the initial magnetite has been largely consumed in the formation of the metamorphic silicates, and a significant portion of the remaining finely disseminated magnetite is commonly poikilitically enclosed within the silicates.

Submember O. In contrast to the adjacent magnetite-poor units, submember O is generally characterized by abundant magnetite-bearing granules, and occasionally by small pebbles and lamellae.

Normally, unit O is persistent and is readily recognized in the cores from all holes cutting this horizon except those from holes 12 and 27E, where units M, N, and O could not be differentiated. Similarly, units N and O could not be separated in core from hole 26. Submember O ranges from 5 feet in hole 33 to 26 feet thick in hole 14; however, on the basis of most cores, it generally averages about 18 feet thick. Cores from holes 1 to 3, 7, 23, and 35 do not extend as far down as unit O.

Granule structures are evident in almost all the cores of the area. In the cores from holes east of hole 20, most of the obvious granules consist of nearly solid masses of fine- to medium-grained magnetite in a medium-grained quartz-silicate matrix. In the cores from holes south and west of Iron Lake the granule structures, though less apparent, are clearly outlined by disseminated fine-grained magnetite enclosed within a matrix of fine-grained cummingtonite that has largely replaced most of the original magnetite. The intervening area essentially constitutes a transitional zone between the two types of relic granule structures. Throughout the submember, granules along with some small pebbles are locally concentrated into lamellae and into closely packed, irregular to lenticular layer-like laminated zones that range from $\frac{1}{4}$ to 12 inches in thickness. Fine-grained magnetite also occurs, generally to a lesser extent, as lamellae and thin laminated zones that locally reach 2 to 4 inches in thickness, particularly in cores from some of the western holes.

The silicate-rich quartzose matrix in which the magnetite is distributed varies greatly, mineralogically, throughout the area. Except for minor differences in the content of calcium-bearing metasomatic mineral assemblages, which depend solely upon proximity to pegmatitic veins, the distribution of fayalite, poikilitic cummingtonite and ferrohypersthene, and even of prismatic to acicular cummingtonite, is nearly the same as that described above for the underlying unit P.

The cores from holes 18 and those lying nearer to the gabbro contact are largely magnetite eulysite, consisting mainly of fine- to medium-grained magnetite, quartz, and fayalite in a granoblastic mosaic that also contains trace amounts of hornblende of uncertain but probably later relative age. In cores from holes 10 and 18, medium-grained poikilitic ferrohypersthene can be found enclosing any of these minerals. Medium-grained hedenbergite has partly replaced the fayalite-bearing assemblages in cores from holes 4, 6, 8, 13, 17, and 18. In most of the cores some medium-grained poikilitic-prismatic cummingtonite locally encloses and has slightly replaced all the above silicates and quartz. Fine-grained prismatic to acicular cummingtonite has developed only to a minor extent, usually along grain boundaries where it has clearly replaced all of the above mentioned minerals. Minor amounts of fine-grained hornblende+cummingtonite accompany the fine-grained cummingtonite in cores from holes 6, 10, 11, 15, 17, and 18, the last of which also contains minor amounts of fine-grained hornblende. Some fine-grained cummingtonite grains contain small pleochroic haloes around minute zircon grains. In cores from holes 10, 18, 26, and 34, some fayalite is altered to brown hisingerite adjacent to thin black veinlets of hisingerite, locally containing traces of magnetite.

Fine-grained prismatic to acicular cummingtonite is increasingly more abundant in cores from holes numbered 19 through 26, where the amphibole has partly to almost wholly replaced the earlier fayalite-bearing assemblages, the roundish remnants of which have produced a distinctive mottling in these cores. These mottled clusters of unreplaced grains range from minute up to 5 mm in diameter and generally diminish in size away from the gabbro. Somewhat similar small remnants of hedenbergite in cores from holes 22 and 25 and of ferrohypersthene in cores from hole 21 also occur within the abundant fine-grained cummingtonite. Minor amounts of hornblende+cummingtonite accompany cummingtonite in the cores from holes 21 and 25. Fayalite is altered to dark brown hisingerite adjacent to the thin veinlets of green-brown hisingerite in core from holes 21 and 26.

In the remaining holes to the west of hole 25, fine-grained acicular cummingtonite is particularly abundant and has left only a few small remnants or "mottled clusters" of fayalite or fayalite-bearing assemblages within the mat of amphibole grains. It seems apparent that the initial content of fayalite decreased from east to west within this submember. Many of the remnants are too small to be positively identified nor can the nature of their replacement by cummingtonite be detected except in thin section. No fayalite remnants were found in the cores from hole 31 and holes west of hole 33. It is probable that the mineral either never developed or did so to such a minor extent that it was completely destroyed by the later development of cummingtonite. Minor amounts of hornblende+cummingtonite accompany fine-grained cummingtonite in cores from holes 30 and 32, the latter also contain trace amounts of oligoclase. Minute amounts of pyrrhotite and chalcopyrite are associated with cummingtonite in core from hole 36. Magnetite-bearing green-black hisingerite veinlets occur and follow the bedding in cores of all the holes from 27 to 34, except hole 31. Many of these veinlets are closely spaced (up to 1 inch apart) and they locally accentuate the layered aspect of the cores. These layers, of course, are not of sedimentary origin but their occurrence was probably controlled by the original bedding. Fayalite adjacent to these veinlets is characteristically altered to dark brown hisingerite.

Submember N. Because of its negligible magnetite content, this thin silicate-rich quartzose unit normally forms a rather distinct horizon in all cores except those from holes 12, 26, 28, 29, and 27E, where the unit could not be recognized with certainty. This unit ranges in thickness from 2 feet in hole 33 to 6½ feet in hole 23 but averages about 3 feet thick in the hole west of Argo Lake and about 5 feet thick in those holes south and east of this lake. Holes 1 to 3, 7, and 35 did not extend into unit N.

Most of the material of unit N consists of fine-grained quartz with minor to locally abundant amounts of silicate. In some of the cores from holes east of Argo Lake, many of the quartz grains reach 1 mm in diameter and in a few specimens up to 3 mm.

Most of the cores from holes east of hole 21 contain moderate to locally abundant amounts of fine- to medium-grained fayalite while those west of this hole contain progressively decreasing amounts. In the easternmost group of the cores fayalite is replaced by only trace to minor amounts of cummingtonite, but cores from hole 17 and from those holes lying farther to the west contain increasingly abundant amounts of cummingtonite that surround smaller and smaller unreplaced clusters of fayalite grains. In cores from holes 25, 26, 27, 30, 34, and 36, very small remnants of fayalite probably represent the westernmost occurrence of the mineral in this unit. Moderate amounts of ferrohypersthene occur locally in cores from holes 19 and 20, where some crystals reach diameters of 3 cm. Minor amounts of medium-grained hedenbergite, present in the cores from holes 4, 5, 6, 13, and 22, are locally abundant in core from holes 18, 19, and 20 where the pyroxene has been partly replaced by fine-grained cummingtonite and some hornblende+cummingtonite. Thin hisingerite veinlets cut some of the cores from holes 17, 21, 23, and 32; fayalite immediately adjacent to these veinlets is generally altered to brown hisingerite.

As mentioned above, cummingtonite is particularly abundant in the cores from holes west of hole 21. Although cummingtonite has clearly replaced fayalite in some of these cores, much of the cummingtonite has been formed directly by the replacement of magnetite-bearing quartzose granule structures. Minor amounts of calcite and traces of actinolite are intimately associated with cummingtonite that has partly replaced granule structures in cores from holes 28, 36, and 37.

Almost every hole has yielded some fine-grained magnetite sparsely distributed through the unit. Although some is disseminated through the silicate-bearing quartzose layers, most of the magnetite is concentrated within a few thin layers up to $\frac{1}{8}$ inch thick or as thinly layered zones up to 1, or rarely 2, inches thick.

Submember M. In contrast to other submembers of the Upper Cherty member the abundant magnetite-rich layers of unit M are, in general, relatively thin and regularly bedded. The contact between units L and M is gradational and difficult to determine in some cores, particularly those from many of the western holes where cummingtonite occurs abundantly in both units. In most of the other cores, however, the presence of a pebbly and granule fabric near the bottom of unit L, and the somewhat greater abundance of silicates — particularly fayalite — in the quartzose layers of unit M, aid in determining this contact.

Unit M is thickest, $35\frac{1}{2}$ feet, in hole 4 and thinnest, $9\frac{1}{2}$ feet, in hole 32, but normally ranges between 15 and 25 feet. Although generally somewhat thinner in cores from the western part of the area, the regional variation is irregular. Unit M was recognized in cores from all drill holes except numbers 1 to 3, 7, and 35, none of which extended down to this horizon.

About half of the bedded taconite-strata of this unit consists largely of fine-grained magnetite occurring as thin individual layers, up to about $\frac{1}{2}$ inch thick, within quartzose material. Nearly equal amounts of bedded taconite-strata consist of similarly thin layers of magnetite alternately interbedded with equally thin layers of locally silicate-rich quartz, which collectively form relatively thick layered zones that range from $\frac{1}{4}$ to 12 inches, although most are 2 to 6 inches thick.

The massive taconite-strata consist of slightly irregularly bedded, locally silicate-rich, quartzose layers that range from $\frac{1}{4}$ to 9 inches, but are most commonly 2 to 4 inches, thick. The quartz is fine-grained except in most of the cores from holes near the gabbro, where some grains reach 1 mm in diameter, but the associated magnetite is rarely appreciably coarsened. As noted in units N and O, there is also a definite transition from east to west with respect to the occurrence and abundance of the various silicates within the quartzose and magnetite-rich strata of the unit.

In essence, fayalite with traces of later cummingtonite is characteristic of the eastern cores, and in the central area numerous small remnants of fayalite and of fayalite and ferrohypersthene are found within increasingly abundant cummingtonite, which finally becomes dominant in the western part of the region. The detailed distribution of the silicates is as follows.

Cores from holes 18, 19, and 20 and holes farther east commonly contain moderate amounts of fine- to medium-grained fayalite which occur mainly within the quartzose layers. Minor amounts of fine-grained ferrohypersthene are also present in almost all of these cores, and in those from holes 12 and 20 a few porphyroblasts reach 3 to 4 cm in diameter. Minor amounts of fine- to medium-grained hedenbergite also occur mainly in the quartzose layers of this eastern group of cores, except those from holes 5, 10, and 17 where the pyroxene is locally abundant. In almost all of these cores fine-grained cummingtonite and other amphiboles are virtually absent, but become increasingly common in cores from holes 18, 19, and from holes farther west. Traces to minor amounts of cummingtonite have replaced fayalite in the cores from holes 18 and 19; however, the amphibole is more commonly found incipiently to largely replaced by the pyroxenes. Core from hole 19 contains coarse-grained ferrohypersthene remnants within fine- to medium-grained cummingtonite and hornblende+cummingtonite, as seen in Figure 14. The finer grained amphi-

FIGURE 14. — Ferrohypersthene remnants in amphibole matrix. In this thin section remnants of coarse-grained ferrohypersthene (dark gray) have been partly replaced by medium- to fine-grained cummingtonite and hornblende+cummingtonite (light gray). Note that the amphiboles are considerably coarser grained near the pyroxene remnants. (19-158; 9X)

boles commonly contain small pleochroic haloes around minute zircon grains. A similar specimen of poikilioblastic ferrohypersthene in core from hole 10 has been partly replaced by medium- to coarse-grained hornblende+cummingtonite intimately associated with small apatite crystals. Moderate amounts of fine- to medium-grained cummingtonite and minute amounts of hornblende+cummingtonite have clearly replaced hedenbergite in cores from holes 5 and 17. Cores from holes 10 and 17 also show hedenbergite that has been incipiently replaced by fine-grained hornblende+cummingtonite and traces of hornblende; trace amounts of apatite are closely associated with these amphiboles. A few thin magnetite-bearing hisingerite veinlets cut some of the cores from holes 5 and 13.

Fine-grained cummingtonite is generally abundant in the cores from holes immediately south and west of Argo Lake. Moderate amounts of small fayalite remnants are found in cores from holes 21, 23, and 24. Figure 15 shows typical corroded remnants of fayalite that have been partly resorbed during the formation of the surrounding granoblastic quartz-

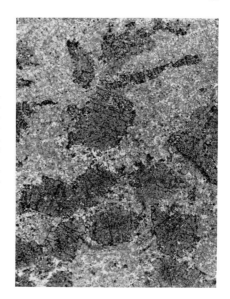

FIGURE 15. — Fayalite remnants in cummingtonite-rich matrix. Typical fabrics are shown by remnants of coarse-grained fayalite (dark gray) with irregular corroded margins, which are locally accentuated by traces of extremely fine-grained magnetite. The remnants occur in a fine-grained granoblastic matrix (light gray) of quartz, cummingtonite, and some hornblende+cummingtonite. (21-325; 9X)

cummingtonite matrix that also contains traces of hornblende+cummingtonite. Minor amounts of medium-grained ferrohypersthene in core from hole 21 and moderate amounts of fine- to medium-grained hedenbergite from holes 23 and 24 also occur as remnants within cummingtonite in this region.

In the remaining holes farther to the west, fine-grained cummingtonite is also locally abundant but less so than in the region just described. A few minute remnants of fayalite, in cores from holes 25, 30, and 34, and of hedenbergite, in cores from holes 30 and 32, are present, although fine-grained cummingtonite is generally the only silicate apparent in these western holes. In thin section, however, it is seen to be commonly associated with minor amounts of actinolite, and in a few places, with plagioclase (oligoclase?). Minute grains of zircon have produced small pleochroic haloes in many of the amphibole grains. Trace amounts of chalcopyrite occur with the cummingtonite- and actinolite-rich parts of the cores from holes 36 and 37. Core from hole 29 contains a few magnetite-bearing hisingerite veinlets.

Relic granule structures are not apparent in the highly recrystallized group of cores from the eastern holes, having been almost completely replaced by cummingtonite in the regions south and west of Argo and Iron Lakes. Numerous quartzose granule structures outlined by fine-grained magnetite, however, are increasingly apparent in the quartzose layers of the cores in the western part of the area because they have been only partly replaced by cummingtonite.

Units M, N, and O could not be recognized as separate units in core from hole 27E, but a variety of relic granule structures is typically abundant in cores from this interval. Most of the granules consist of different amounts of fine-grained quartz, magnetite, and minnesotaite. Many of these granules contain small magnetite euhedra, and a few contain finely disseminated hematite. Minor amounts of ankerite are associated with minnesotaite in some places and a few minnesotaite veinlets cut the granules. One interesting specimen containing stilpnomelane-rich quartzose granules cut by numerous veinlets of fine-grained minnesotaite has been cut in turn by a thin stilpnomelane-quartz-siderite vein.

Submember L. This unit is similar in many aspects to the adjacent submembers. Although slightly pebbly throughout, it is generally somewhat less conglomeratic than the overlying unit K whereas the underlying unit M is distinctly less pebbly. In the western part of the area the upper part of the unit generally contains relatively numerous thin magnetite layers and the bottom part numerous densely packed magnetite-rich granules. This transition from abundant thin magnetite layers to the abundant magnetite-rich granules is not as apparent, however, in most of the cores from the eastern holes as in those from the western holes.

For the purposes of discussion of submembers L and K only, holes numbered from 4 to 20 are called the eastern holes. Unit L averages about 15 feet in thickness in the far eastern holes and about 45 feet in the far western, and is 46 feet thick near Mesaba. There is no uniform gradation of thickness from the thickest, 51 feet in hole 21, to thinnest, 8 feet in hole 13, but 30 feet can be used as a crude average thickness for the unit. Units K and L could not be separated in cores from holes 5 and 34, where combined thickness of these units was 43 and 84 feet respectively. Cores from holes 1 to 3, 7, and 35 did not extend to L.

Some fine-grained magnetite occurs as single layers up to $1/4$ inch thick, although a few reach $3/8$ inch thick. Most of the bedded taconite-strata, however, consist of many closely spaced thin layers of magnetite — generally less than $1/8$ inch thick — that collectively form layered zones ranging from $1/4$ to 6 inches, and, rarely, to 12 inches, thick. Most of these layered zones are between 2 and 3 inches thick and can be easily recognized in cores from throughout the area. The thicknesses and frequency of occurrence of the layered magnetite-rich zones generally diminish with depth and the bottom part of the unit is commonly sparsely layered, containing numerous magnetite-rich granule structures. The thin layers interbedded with the magnetite in the layered zones generally consist of fine- to medium-grained quartz that locally contains various silicates. In cores from the eastern holes, the magnetite-rich zones commonly contain locally abundant medium- to coarse-grained ferrohypersthene. Medium-grained hornblende is also locally abundant and moderate quantities of actinolite occur in cores from holes 6 and 18. Thin layers of fine-grained cummingtonite constitute almost all the material between the closely-spaced magnetite layers in the cores from the western holes. Minor amounts of actinolite accompany cummingtonite in some of these cores and trace quantities of calcite occur in the cummingtonite-bearing core from hole 31.

The silicate-bearing quartzose massive taconite-strata in both the eastern and western cores are slightly lenticular and range in thickness from ¼ to 8 inches, although most are between 2 to 4 inches. Although magnetite-rich granules are commonly present throughout, they become increasingly apparent and abundant with depth, except in most of the eastern holes where slight recrystallization of the taconite nearer the gabbro has apparently destroyed the granules, presumably once present. The massive quartzose layers separating the thinly bedded magnetite-silicate-rich zones in the cores of all the eastern holes typically contain moderate amounts of fine- to medium-grained ferrohypersthene as well as some exceptionally coarse-grained porphyroblasts as in cores from holes 5, 18, and 20, where diameters of 3, 2.5, and 4 cm were attained, respectively. Moderate quantities of medium-grained hedenbergite are also associated with ferrohypersthene in the quartzose layers in many holes and it is locally abundant in core from hole 17. Pegmatitic vein material, mainly aggregates of feldspar-quartz-hornblende, occurs in cores from holes 5, 12, 18, and 20.

The quartzose massive taconite-strata of the cores from the western holes locally contain abundant quantities of fine-grained cummingtonite. A great variety of relics of fine-grained magnetite-quartz granules in these beds can be found in all stages of progressive replacement by fine-grained cummingtonite, as illustrated in Figure 16. In these cores, cummingtonite is commonly intimately associated with minor to trace amounts of very fine-grained greenish biotite, plagioclase (oligoclase?), chalcopyrite, and minute grains of zircon that have produced small pleochroic haloes within the cummingtonite. Moderate amounts of fine- to medium-grained hedenbergite, present in the quartzose layers of the cores

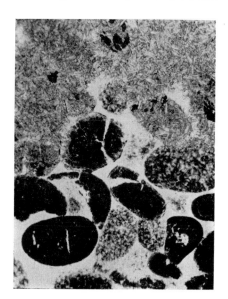

FIGURE 16. — Relic magnetite-quartz granules partly replaced by cummingtonite. Fine-grained quartzose granules, containing "dusty" magnetite, in nearly all stages of reconstitution into fine-grained cummingtonite can be seen. Trace amounts of fine-grained biotite and plagioclase commonly accompany cummingtonite during metasomatic replacement in this material. (32-260; 30X)

from many of the western holes, occur in cores from holes 21 and 28 as coarse-grained porphyroblasts reaching 1 and 2 cm, respectively. A thin section of this material from hole 21 shows that the hedenbergite (and some actinolite) has partly replaced some magnetite-quartz granules. The pyroxene, which has been selectively and slightly altered to nontronite, exists as distinct corroded remnants within a matrix of fine-grained cummingtonite and granoblastic quartz. Many poorly preserved granule structures, outlined by finely disseminated magnetite, are still recognizable within the silicate-rich zones of the cores. Some clusters of medium-grained ferrohypersthene occur in the quartzose layers of core from hole 28, and a few coarse-grained porphyroblasts — ranging to 2.3 cm in diameter — are associated with the fine-grained hedenbergite and calcite in core from hole 37. Upon closer inspection, ferrohypersthene from these cores is found to exist as remnants within, or as partly replaced by, cummingtonite and hornblende+cummingtonite, exactly as illustrated in Figure 14. Introduced calcite has partly replaced magnetite-quartz granules and their quartz matrix in core from hole 31, as seen in Figure 17. Here a network of the carbonate surrounding several granules can constitute a

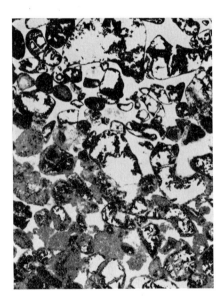

FIGURE 17. — Relic granule structures. Incipient recrystallization (top) frequently accompanies the introduction of traces of calcite (gray, bottom) into quartzose taconite. In this specimen a lace-like network of coarse-grained calcite locally surrounds, and has partly replaced, fine-grained magnetite-quartz granule structures that have been slightly modified during recrystallization. (31-97; 9X)

single large lace-like crystal, all of whose parts are in optical continuity. Minute platy crystals of chlorite(?), or possibly minnesotaite(?), are commonly associated with the calcite. Thin hisingerite veinlets, apparently along fractures, cut the core from hole 25, but have had little if any effect upon the earlier silicates.

In core from hole 27E near Mesaba, unit L, difficult to distinguish from those adjacent to it, was differentiated on the basis of a greater overall abundance of granules and the somewhat persistent occurrence of moderately thick, layered magnetite-rich zones. These layered zones are mostly about $\frac{1}{2}$ to 2 inches thick and in some places contain many closely packed layer-like accumulations of magnetite-rich granules. The bedded parts of the unit diminish in abundance rather gradually with depth. The magnetite-rich strata of the core are generally separated by massive fine-grained quartzose of silicate-rich quartzose layers about 2 to 3 inches thick. Most of the granule structures within the quartzose layers are largely composed of very fine-grained minnesotaite, magnetite, and chert. Some of the darker granules also contain fine-grained stilpnomelane, especially near the thin stilpnomelane- and carbonate-bearing magnetite-rich layers of the core. Moderate quantities of calcite locally occur as irregularly distributed masses throughout the core and are associated with the above-mentioned minerals.

Submember K. Unit K is heterogeneous with respect to the variety of relic sedimentary bedding structures it contains. Magnetite-rich strata are plentiful but show a wide variation in perfection of bedding and of association with the silicates occurring in the unit. Generally, unit K is also somewhat conglomeratic and pebbly, and the massive quartzose layers normally have a distinct granule texture. Although greatly oversimplified it can be said that the magnetite-rich strata and the interbedded quartzose beds are thin and closely spaced near the top of the unit and are commonly somewhat thicker and more widely separated near the bottom.

Unit K could not be clearly separated from unit L in cores from holes 5 and 34, or from unit J in core from hole 8; in these cores the combined thicknesses of the respective groups of units is 43, 84, and 41 feet. It was recognized in cores from every other hole, except holes 1 to 3, 7, 15, 17, and 35, all of which did not extend to this unit. The unit is thickest, $47\frac{1}{2}$ feet, in hole 27E near Mesaba, and thinnest, 19 feet, in hole 4. There is an irregular decrease in average thickness from about 40 feet in the western area to about 30 feet in the area of the eastern holes. For the purposes of the discussion of this unit, holes numbered from 4 to 20 are again called the eastern holes.

Two varieties of magnetite-rich bedded taconite-strata occur in this unit, both of which appear to be equally abundant and equally distributed throughout. In one instance, magnetite occurs as single thin layers that commonly range in thickness up to about $\frac{3}{8}$ inch. These layers, generally somewhat lenticular or irregularly bedded, commonly diverge into a series of thinner or thicker beds separated by a few thin lenticular quartzose- to silicate-rich layers. The other occurrence is found within thicker interbedded zones consisting of thin magnetite layers, of the type just described, intercalated with similarly thin quartzose- or silicate-rich layers. These layered zones range from $\frac{1}{4}$ to 8 inches, though most are from 1 to 3 inches thick. In most of the cores the magnetite layers, whether occurring singly or within the layered zones, appear to be slightly thinner, on the average about $\frac{1}{16}$ to $\frac{1}{8}$ inch nearer the top and slightly thicker near the bottom, about $\frac{1}{8}$ to $\frac{1}{4}$ inch, although unusually thin and thick magnetite layers can occur irregularly distributed throughout cores of the western holes. This distribution of magnetite layers is not apparent in most of the cores from the eastern holes, where the unit is mainly delineated by recognizing the units above and below. In many of the cores from the entire district small irregular patches of closely packed magnetite-rich granules are found among the magnetite layers. The locally silicate-bearing, quartzose, massive taconite-strata that separate the magnetic bedded taconite-strata in the district are normally somewhat lenticular and generally have a granule fabric, particularly in the western part of the area. These quartzose layers, ranging from $1/8$ to 9 inches, are mainly between 1 and 4 inches thick. In a very general way these massive strata are also thinner near the top of the unit and thicker near the bottom. The areal distribution of silicates within both the bedded and massive taconite-strata is described as follows.

In the eastern cores the massive taconite-strata consist largely of fine- to medium-grained granoblastic quartz, with some grains up to 2 mm in diameter in most cores. Small magnetite-rich pebbles are easily recognized within these layers; distinct granules are uncommon — presumably having been destroyed by recrystallization. These structures may be seen, however, in thin section, where they are outlined by sparsely disseminated magnetite in quartz. Minor amounts of silicates occur only locally in the quartzose layers, and in most of these places the silicates are those that extend outward from the magnetite-rich parts of the cores into the quartzose layers.

Medium-grained actinolite is locally abundant within and adjacent to the magnetite-rich strata in almost all the cores from the eastern holes and moderate quantities of medium-grained cummingtonite accompany actinolite. In cores from holes 18 to 20 the amphiboles are equally abundant. Except immediately adjacent to the magnetite-rich strata, only minor amounts of these amphiboles are present in the quartzose layers. Medium-grained anhedra of ferrohypersthene, incipiently replaced mainly by hornblende+cummingtonite, occur within the layered magnetite-rich zones, and to a lesser extent in the adjacent quartzose layers in cores from holes 5, 6, 9, and 13. Coarse-grained anhedra of this pyroxene, up to 15 mm in diameter, occur as remnants within a fine-grained matrix of hornblende+cummingtonite in cores from hole 20. Coarse-grained anhedra of hornblende occur locally in the layered magnetite zones of the cores from holes 5 and 12, and also from hole 20 where it is associated with feldspar. The feldspar-hornblende-bearing pegmatitic veinlets associated with this metasomatic amphibole, however, were found only in cores from hole 12. Thin veinlets of hisingerite locally cut the core from hole 6.

Fine-grained cummingtonite is generally abundant within the layered magnetite-rich zones of cores from the western holes. If the individual layers of magnetite were thinner and more regularly bedded, the layered zones within this unit would resemble the uniformly laminated, magnetite-cummingtonite-rich zones of units F and G, described below. Fine-grained actinolite and in several places hornblende+cummingtonite are commonly also present among the magnetite-rich strata of the western cores, but these amphiboles are normally subordinate in amount to cummingtonite. Minor amounts of feldspar are associated with the amphiboles in core from hole 21. Magnetite-quartz granules adjacent to the amphibole-magnetite-rich layers are partly to wholly replaced by cummingtonite and (or) actinolite in a manner similar to that illustrated in Figure 21.

The massive taconite-strata in the cores from the western holes consist largely of fine-grained quartz, locally abundant cummingtonite, and in a few places, some actinolite. Cummingtonite, found throughout the unit, is commonly more abundant in the upper part. In the western cores, the presence of cummingtonite helps distinguish this unit from the cummingtonite-poor overlying unit J. The wide variety of relics of primary granule structures is the most conspicuous feature of the more quartzose layers. Most of these dusty-textured granules, composed of minute micron-sized magnetite grains within very fine-grained chert-like quartz, contain many syneresis-like cracks. Granule structures in all stages of destruction, as a result of reconstitution of the magnetite and quartz into cummingtonite, are clearly visible in all the western cores. In many places only incipient replacement by a few small cummingtonite prisms has taken place within the granules, or across the granule boundaries into the quartz matrix, or even entirely within the surrounding quartz matrix. In numerous other places the rock has been almost entirely reconstituted into a decussate matrix of fine-grained cummingtonite in which the granule structures are largely obliterated. Many intermediate stages of replacement of granules by fine-grained cummingtonite can be observed in Figure 18. This figure shows fabrics typical of those found in the western cores. A similar but much less common relationship is found among the relic magnetite-quartz granules that have been partly replaced by actinolite and calcite. Actinolite in these places is commonly slightly altered to nontronite. Coarse-grained hedenbergite, occurring in some quartzose layers of cores from holes 21 and 28, has been partly replaced by cummingtonite and hornblende+cummingtonite. Large remnants of ferrohypersthene anhedra, up to 10 mm in diameter, are found within a matrix of cummingtonite and hornblende+cummingtonite in core from hole 22; several granule-like structures consisting of finely disseminated magnetite are preserved within ferrohypersthene grains. Thin hisingerite veinlets cut cores from holes 21 and 28, and minor amounts of nontronite have developed in the adjacent silicates.

In the westernmost hole, 27E near Mesaba, the distribution of the dark bedded taconite-strata is similar to that noted in the cores to the east; these layers consist largely of fine-grained magnetite, quartz, and some ankerite. The massive taconite-strata are largely very fine-grained chert-like quartz with locally abundant very fine-grained greenish min-

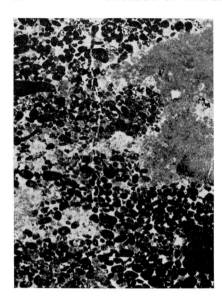

FIGURE 18. — Incipient development of cummingtonite. The quartzose layers of *layered (magnetite) quartz taconite* commonly contain numerous relic "dusty" magnetite granules, although in many places they have been largely replaced by fine-grained cummingtonite (light and dark gray). The picture shows typical granules that have been incipiently replaced by cummingtonite and traces of actinolite. (32-238; 4X)

nesotaite needles as well as disseminated subhedral magnetite crystals that have poorly perserved the outlines of former granule structures. In many places these granule structures are visible, but where minnesotaite has developed in abundance they have been largely destroyed. Some specimens are cut by quartz-stilpnomelane-ankerite veinlets and minor amounts of fine-grained ankerite and stilpnomelane have developed in the adjacent magnetite-rich and quartzose beds of the cores.

In the discussion of the remaining submembers of the formation, holes numbered from 25 to 37 are again considered as the western holes; hole 21 and those lying east of it are called the eastern holes, as before.

Submember J. Unit J contains a variety of magnetite-rich, relic sedimentary structures. In the cores from all the western holes and holes 18 and 21 of the eastern group, the unit could be subdivided even further into three more or less distinct subunits. In these cores the bottom portion of the unit contains abundant lenticular and slightly irregularly-bedded magnetite-rich layers averaging about ½ inch thick. The middle portion is characterized by the almost ubiquitous magnetite-rich pebbles and somewhat coarse fragments of intraformational conglomerates. The upper part typically contains numerous closely-packed magnetite-rich granules. This characteristic distribution is not present in the cores from most of the eastern holes. There, the entire unit contains mainly granules and pebbles throughout and the presence of lenticular, conglomeratic, and brecciated strata is common. In both the eastern and western cores the matrix in which the magnetite is distributed is distinctly quartzose everywhere, although silicates occur locally as described below.

The unit was recognized in all cores, except those from holes 1 to 3, 7, 14 to 17, and 24, all of which did not reach this stratigraphic horizon. It is thickest, 29 feet, in hole 10 and thinnest, 9½ feet, in hole 11, but averages about 17 feet in thickness. The unit is 22 feet thick near Mesaba, where the upper 6 feet of core resembled the upper portions of the western cores. Unit J could not be clearly recognized in core from hole 8, where it was included with unit K to give a combined thickness of 41 feet. Similarly, units I and J were combined in the logging of core from hole 9 to give an over-all thickness of 19½ feet.

In cores from holes 18, 21, and the western holes, the bottom portion of the unit, averaging 7 feet thick, is characterized by the occurrence of relatively thick layers of fine-grained magnetite. These layers are slightly irregularly bedded to lenticular, range in thickness from ⅛ to 2 inches, but are generally about ½ inch thick. In many cores closely-spaced magnetite lamellae form layered zones ½ to 12 inches thick, although most are 2 to 3 inches thick. The magnetite-rich strata are normally separated by 1 to 4 inch thick quartzose layers in which relic granule structures are evident in all this group of cores.

Coarse-grained hedenbergite occurs to a moderate extent in cores from holes 29, 32, 36 (all three of which lack obvious cummingtonite), and 21, and only to a minor extent in cores from holes 27, 28, 30, and 37. It is also intimately associated with coarse-grained ferrohypersthene in cores from holes 18, 21, and 36. In addition, core from hole 21 contains remnants of medium-grained fayalite and ferrohypersthene in a granoblastic mosaic of hedenbergite, quartz, and magnetite along with some cummingtonite and hornblende+cummingtonite; these amphiboles have locally incipiently replaced the other silicates and quartz. A few small striated prismatic grains of loellingite occur interstitially to or within hedenbergite in cores from holes 32 and 36.

Fine-grained cummingtonite, locally abundant in cores from holes 21, 25, 27, 30, 31, and 37, is only moderately abundant in cores from holes 28 and 33. It occurs almost exclusively within the quartzose layers where it has partly replaced the above mentioned silicates and in many places has locally replaced and partly destroyed the magnetite-quartz granules.

The middle portion of unit J, about 6 feet thick on the average, is recognized in the western group of holes on the basis of the normally ubiquitous presence of fine-grained magnetite occurring in the form of pebbles, ranging from 2 mm to about 20 mm, within a fine-grained quartz matrix. Many of the smaller pebbles are similar to the granules and might have had a similar origin. Some of the large pebbles are probably related to intraformation conglomerates, although normally more rounded than the subangular to subrounded tabular fragments typical of these conglomerates elsewhere in the formation. Many cores also contain some magnetite as lamellae and thin lenticular layers that average about ¼ inch thick. In some places closely-spaced lamellae form a few thin laminated zones up to ¾ inch thick. The magnetite-rich strata of the cores are normally separated by ¼ to 8 inch thick layers of the pebbly quartzose material described above.

Minor amounts of fine-grained hedenbergite have partly replaced quartz and magnetite in cores from holes 18, 21, 25, 27, 28, 32, and 33, and coarse grains of it, as stubby prisms up to 5 mm long, occur similarly in cores from holes 29 and 36. Coarse-grained ferrohypersthene is intimately associated with hedenbergite in core from hole 36.

Fine-grained cummingtonite is locally abundant in the quartzose layers of the cores from holes 30 and 32, and to a lesser extent from holes 28 and 29, where it has largely replaced the magnetite-quartz granules of the layers. In these places the pebbly fabric of the taconite is accentuated by the presence of light-colored cummingtonite in the quartzose matrix. Minor amounts of fine-grained actinolite were noted in cores from holes 29 and 30.

On an average the upper 4 feet of unit J in the western group of cores is characterized by numerous closely-packed magnetite-rich granules in a quartzose matrix. In some places the granules are so abundant that they form indistinct layer-like concentrations ranging from $\frac{1}{2}$ to 8 inches thick. A few thin lenticular to irregular layers of fine-grained magnetite, averaging $\frac{1}{4}$ to $\frac{1}{8}$ inch thick, occur in many cores. Minor numbers of small magnetite-rich pebble like those described above are also common in this upper portion of unit J.

Minor amounts of fine- to medium-grained hedenbergite have replaced quartz and magnetite in cores from holes 18, 21, 25, 29, 32, 33, and 35. Fine-grained cummingtonite is locally abundant in core from hole 30, and to a minor extent from hole 32, but was not observed in the remainder of the holes.

A rare occurrence of the blue soda-rich amphibole, riebeckite, is found in core from hole 28, in the form of numerous small acicular to prismatic crystals, up to 2 mm long, bordering some of the magnetite-rich parts of the core. Most of the magnetite has been recrystallized and riebeckite has partly replaced magnetite and quartz throughout the specimen. This replacement feature is most obvious where prismatic grains of riebeckite have formed within zones containing numerous relic dusty-magnetite quartzose granules and have partly destroyed these earlier structures (Fig. 19).

Cores of unit J from the eastern group of holes (less holes 18 and 21) contain a variety of magnetite-rich relic sedimentary structures similar to those just described from the western cores. However, the densely packed granule zones, the conglomeratic zones, and the thick- and thin-layered zones assume no consistent distribution and, consequently, cannot be further subdivided. In contrast to the adjacent units in these cores, the entire unit is more pebbly and rich in granule structures and the presence of brecciated, lenticular, and conglomeratic strata is common. The material separating the various magnetite-rich parts of the cores consists of quartzose layers from $\frac{1}{4}$ to 8 inches thick, averaging 2 to 4 inches. Thin sections of these quartzose layers are seen to contain numerous dusty magnetite granules; many of the granules have been recrystallized,

FIGURE 19. — Riebeckite. Very fine-grained magnetite-quartz granules (dark gray) have been replaced by fine prismatic grains of riebeckite (light gray), which are accompanied by traces of muscovite elsewhere in this and other specimens. Although many of the relic granules are well preserved, there has been some crystallization of the "dusty" magnetite into small anhedra (black). (28-54; 100X)

however, and the granule structures are now outlined by fine-grained magnetite.

Medium- to coarse-grained ferrohypersthene is locally abundant in core from hole 12; only minor amounts of it occur in cores from holes 5, 6, 19, and 23. Moderate quantities of medium- to coarse-grained hedenbergite, which has partly replaced magnetite and quartz, occur locally in nearly all cores from the eastern group of holes. A few grains of fayalite occur as remnants in hedenbergite in core from hole 19. Except for a few minute traces, cummingtonite is virtually absent in the cores of the eastern holes.

Core of unit J from the hole near Mesaba vaguely resembles those from the eastern group of holes just described, although not particularly pebbly nor locally thick-layered with magnetite. Most of the obvious granule structures are magnetite-rich although many of these structures have been replaced by fine-grained minnesotaite. The core from this hole is locally brecciated (a primary feature) and contains several crenulated lamellae and lenticular beds as well as several stylolites that are marked by abundant small magnetite grains and needles of brown stilpnomelane. The immediately adjacent rock, generally quartzose, also contains abundant amounts of pale green minnesotaite and moderate amounts of brown and some green stilpnomelane needles, small disseminated magnetite grains, and irregular siderite rhombohedrons. It is significant that some veinlets of minnesotaite cut the stylolite structures.

Submember I. This unit is defined by the presence of algal structures. Grout and Broderick (1919, p. 21) were among the first to recognize the organic origin of these structures. Although some algal colonies were little disturbed after their formation and have been recovered intact in some cores, more commonly only conglomeratic zones containing fragments of the algal structures define the unit in cores from the Eastern Mesabi district. The intraformational conglomeratic nature of the unit is persistent throughout the district; in some cores where the algal nature of the fragments is less clear, the unit can still be tentatively delineated on this basis. Relic layering, generally as indistinctly bedded lamellae, rarely appears in these rocks.

Submember I was recognized in cores from all the drill holes of the district, except 1 to 3, 7, 13 to 17, 23, 24, 31, and 37, all of which terminated before reaching this unit. The algal unit is thickest, 8 feet, in hole 20, and thinnest, 2½ feet, in hole 21; in most of the other holes the unit ranges from 4 to 6 feet thick. In the logging of core from hole 9, unit J was included with the algal horizon to give a combined thickness of 19½ feet.

Numerous algal and granule structures are outlined by disseminated fine-grained magnetite within poorly bedded to massive fine-grained quartz. Commonly, both magnetite and quartz have been slightly recrystallized in cores from near the gabbro, but many relic granule and algal structures are still apparent. The granule and algal structures in many of the cores from holes south and west of Iron Lake consist in part of hematite (Fig. 20). Most of the magnetite in these structures has been recrystallized and subsequently altered to martite. Many of the granules

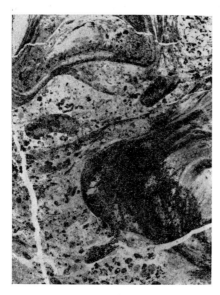

FIGURE 20. — Algal structures. The preservation of excellent algal structures is common even in slightly recrystallized and oxidized taconites. In this specimen the algal structures are defined by fine-grained hematite (black), probably martite, and some slightly recrystallized parts of these structures consist of subhedral martite grains. The surrounding quartzose matrix contains numerous slightly recrystallized fine-grained martite-quartz granules. (30-111; 6X)

and some parts of the algal structures contain dusty hematite, which may, or may not, be martite. Hematite is locally abundant in cores from holes 25 and 27, occurs in moderate quantities in cores from holes 28, 33, and 36, but is present only to a minor extent in cores from holes 22, 29, 32, 35, and 27E. It is absent, or possibly present in negligible amounts, in all other cores of the unit. The dusty hematite is almost invariably associated with minute amounts of calcite that is interstitial to the fine-grained quartzose matrix. Traces of actinolite accompany calcite where these minerals have incipiently replaced algal and granule structures.

The relatively few lamellae and the numerous small pebbles ranging up to 3/4 inch in diameter that occur locally within the unit consist of fine-grained magnetite disseminated within a fine-grained quartz matrix. In cores nearer the gabbro, magnetite becomes medium grained; in core from hole 8, some imperfect octahedra of magnetite measure about 2 mm along an edge. Quartz is generally fine-grained in most cores, but in some from near the gabbro a few grains reach 5 to 7 mm in diameter.

Minor quantities of medium-grained hedenbergite occur in most of the cores from the eastern holes as well as from holes 25, 32, and 33. Medium- to coarse-grained anhedra of andradite, occurring as a few small clusters and lenses up to 2 inches thick, are found in the cores from holes 22, 25, and 28. Minute amounts of calcite are commonly present with hedenbergite and andradite. Cores from holes 28, 30, and 35 contain minor amounts of fine- to medium-grained clinohypersthene. This uncommon monoclinic pyroxene occurs with oligoclase and hedenbergite to form small irregular granoblastic masses within a mosaic of quartz and magnetite. Minute amounts of actinolite and hornblende+cummingtonite occur interstitially, or have partly replaced the pyroxenes, along with traces of calcite.

Except for very fine-grained minor quantities in core from hole 29, cummingtonite is rarely found in unit I. Some dusty magnetite quartz granules in core from hole 35 consist of partly replaced prisms of riebeckite, in a manner similar to that illustrated in Figure 19, where riebeckite is accompanied by traces of interstitial muscovite.

In core from hole 27E, algal structures are largely defined by a cloudy material, somewhat resembling allophane, that is mostly interstitial to the fine-grained quartz matrix of the rock. Some parts of the core have been recrystallized into coarse-grained quartz and parts of the relic algal structures are largely destroyed in these areas. In other parts of the core, small ankerite rhombohedrons accompanied by minor numbers of fine stilpnomelane and minnesotaite needles have developed and partly replaced these structures.

Submember H. This unit is characterized, in part, by the presence of yellowish to dark greenish amphibole assemblages that occur as thin fine-grained layers among the magnetite-rich layered zones of the rock. Drill holes 1 to 3, 7, 13 to 17, 23, 24, 29, 31, and 37 did not extend into unit H, but the unit can be easily recognized in all other drill cores. It ranges in thickness from 6 feet, in cores from holes 33 and 35, up to 15 1/2 feet in core from hole 5. In general, however, it averages about 8 feet thick in the western holes and gradually increases to an average of about 12 feet in the eastern holes. The unit correlated with unit H in the core from the hole near Mesaba (27E) is nearly 30 feet thick.

The bedded taconite-strata of the unit consist largely of fine-grained magnetite occurring as slightly irregularly-bedded to lenticular lamellae and thin layers up to ¼ inch thick. These magnetite-rich strata most generally occur as groups of closely-spaced beds, normally interlayered with, and enveloped by, actinolite- and hornblende-rich layers of similar thickness. These layered zones of magnetite and amphibole, ranging from ½ to 6 inches, are mainly about 2 to 3 inches thick. Most specimens have been slightly recrystallized and the probably original lamellae are now seen in thin section as thin bands of closely packed small magnetite anhedra. Actinolite and hornblende have partly replaced some of the relic layering in several of the cores. Numerous distinct relic granule structures are commonly preserved by clusters of small magnetite anhedra within the fine-grained quartz layers of the rock. The most significant feature seen in thin sections is the partial to complete replacement by hornblende and actinolite of the quartz-magnetite granule structures adjacent to the magnetite-amphibole layered zones, as shown in Figure 21. Locally abundant in core from hole 21, cummingtonite also occurs in minor amounts in cores from many other holes. In these specimens, cummingtonite is

FIGURE 21. — Granule structures replaced by metasomatic hornblende. Many of the laminated beds (gray and black) of this unit served as channel-ways for metasomatizing solutions. Locally these solutions permeated the surrounding layers, where they have largely reconstituted the adjacent magnetite-quartz granules into granule-like clusters (dark gray) of hornblende and actinolite grains. Outward, beyond this zone of replacement, the granules (black) consist of slightly recrystallized magnetite and quartz. Farther from the pegmatites, similar fabrics are shown by cummingtonite. Adjacent to some of the thicker and richer calcium-rich amphibole and magnetite lamellae, cummingtonite-bearing granules also appear between the hornblende-rich and magnetite-rich granules. (32-177½; 5X)

intimately associated and contemporaneous with the calcium-rich amphiboles, and the entire silicate assemblage commonly assumes a light green color, depending on the mixture present. Farther from the source of metasomatizing solutions, cummingtonite would be found occurring alone.

The massive taconite-strata that separate the magnetite-amphibole-rich strata of the unit consists of fine-grained quartzose layers ranging from about ½ to 6 inches, but mostly 1 to 4 inches, thick. Minor amounts of dusty and fine-grained magnetite occur in relic granule structures, which are common in the quartzose layers throughout the area. Most of these structures have been largely destroyed by recrystallization in many cores from near the gabbro. In numerous other places granules have been largely replaced by amphiboles, as discussed above. Dusty hematite (martite?), which occurs in some relic quartzose granules in core from hole 36, is rare.

Fine-grained fayalite anhedra are present in minor quantities in cores from holes 4, 6, 12, 18, and 21, where it generally appears within the quartzose layers. Several coarse grains of ferrohypersthene, up to 3 mm in diameter, occur within the magnetite-rich strata of the cores from holes 12, 18, and 21. Moderate quantities of medium-grained hedenbergite are present, mainly within the quartzose layers, in cores from holes 4, 18, and 20; only minor amounts are found in cores from holes 5, 19, 22, 25, and 36. All the anhydrous silicates mentioned here are commonly incipiently replaced by, and a few even occur as remnants within, hornblende and actinolite.

Core from the westernmost hole, 27E, contains magnetite-rich layered zones that normally average about one inch thick, and the quartzose layers normally average about 2 to 3 inches thick. Fine-grained minnesotaite is common among the magnetite-rich strata of the unit, and magnetite-quartz granule structures are common in the quartzose layers as noted in the eastern cores. In several places these structures and their surrounding quartzose matrix have been entirely replaced by calcite while still perfectly preserving the granule structures, which now consist of dusty martite within medium-grained calcite, as shown in Figure 29.

Upper Slaty Member

The Upper Slaty member has been divided into seven stratigraphic submembers: A to G. Generally, in passing from the Upper Cherty into the Upper Slaty member, there is a gradational increase in the well-bedded nature of the rocks, largely because the thinly bedded zones become thicker, more regularly bedded, and numerous. Except for a usual paucity of fayalite, most of the mineral assemblages and grain fabrics are similar to those found in the Upper Cherty member.

Parts of the Upper Slaty member were recognized in the cores from holes 2 to 7, 18 to 22, 25 to 28, 30, 32 to 35, and 27E. The entire section of the member was recovered in cores from holes 18, 21, 25, and 32, but also a nearly complete section can be recognized in cores from holes 5, 30, 33, and 34. This member normally shows a uniform thickness of about 120 feet.

Although the member perhaps contains more thinly bedded magnetite-bearing strata, the average magnetite content is normally slightly less than that of the Upper Cherty member. Generally, intimate association with silicates and the finely disseminated nature of the magnetite prevent the current utilization of many of the units of this member, but recent advances in the beneficiation of taconite might rectify this condition in the near future.

Submember G. The presence of andradite is locally characteristic of unit G, which consists largely of massive granule-textured quartz taconite. The unit contains many thin, laminated, magnetite-rich zones; in comparison with the adjacent submembers of the formation, however, these beds are thinner and more sparsely distributed.

Cores of unit G were recovered from holes 4 to 6, 18 to 20, 25 to 28, 30, 32 to 35, and 27E, but units F and G were combined in the logging of the cores from holes 19, 20, and 27E (see discussion of submember F below). The average thickness of unit G is about 25 feet, although thickness ranged from $19\frac{1}{2}$ feet in hole 34 to $29\frac{1}{2}$ feet in hole 28. The basal parts of the cores, termed unit G during logging, from holes 4, 5, 6, 12, and 22, contain rock types identical to those described below for the overlying submember F. The thickness of unit G in each of these five holes is 11, 18, $19\frac{1}{2}$, $15\frac{1}{2}$, and 9 feet, respectively.

Although many of its thin laminated beds and zones of closely-packed granules are rich in magnetite, the unit consists largely of granule-bearing quartzose massive taconite strata. Though ranging from $\frac{1}{2}$ to 12 inches, most of these quartzose layers are between 1 and 8 inches thick. Slightly pebbly horizons, consisting mainly of magnetite-bearing rounded fragments of intraformational origin, occur near the bottom of the unit in cores only from holes 5, 6, 12, 20, and 27; however, it is questionable if these beds can be correlated. The quartz of these layers, though generally fine-grained, is locally medium-grained in cores from hole 18 and from those farther east, nearer the gabbro.

The granule-rich nature of the quartzose layers is rather obscure in cores from holes east of hole 21. Its granule fabric becomes increasingly obvious in cores from holes farther west, a feature also accompanied by a corresponding decrease in abundance of magnetite-rich lamellae in the western area. Under the microscope many of the quartzose layers in cores from the eastern holes are found to contain fine-grained quartzose granules, dusted with very finely disseminated magnetite grains ranging from 1 to 20 microns in diameter. In numerous other layers, however, some recrystallization occurred, and the still recognizable relic fine-grained quartzose granule structures are now outlined by fine-grained magnetite. Most of the granule structures in cores from the western holes consist of fine-grained quartzose granules dusted with very finely disseminated magnetite and some hematite (martite?) as illustrated in Figure 22. Numerous zones of slight recrystallization appear to be somewhat randomly distributed throughout the western area, probably the result of proximity to pegmatitic veins. Within these recrystallized zones, granule structures are outlined by fine-grained magnetite in quartzose material that commonly contains minor quantities of actinolite and calcite, and in some places traces of plagioclase, which is now largely altered to white mica (sericite?); some actinolite has been altered to nontronite.

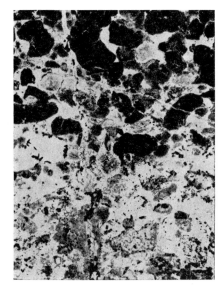

FIGURE 22. — Relic granule structures. Here one of the many results of modification of granule structures can be seen where delicately preserved structures (dark area) of extremely fine-grained magnetite and some hematite (martite) in fine-grained quartzose material, have been only slightly recrystallized (light area) but the relic structures are now vaguely outlined by fine magnetite grains. Minor amounts of calcite, actinolite, and plagioclase occur in the recrystallized zone and traces of calcite and actinolite also occur among the incipiently recrystallized, dark, granule-rich parts of the specimen, particularly adjacent to the quartzose veinlets. (35-44; 9X)

Bedded taconite-strata of the unit consist of thin, magnetite-rich, laminated zones that range in thickness from a fraction of an inch up to an exceptional 12 inches, but most are 1 to 3 inches thick. The magnetite is normally fine-grained and the material between the magnetite lamellae is generally quartzose, although in many places some silicates have developed. The laminated zones are somewhat similar in appearance to those in the overlying unit F, but they are generally more irregularly bedded and thinner. The layering of the unit becomes obscure and less abundant in cores from holes west of hole 21 and very poorly layered, densely packed zones of granules rich in magnetite become prominent. These irregular — presumably lenticular or tabular — zones of abundant granules generally range in thickness from $\frac{1}{2}$ to 8 inches, but the majority are 2 to 3 inches or less.

Moderate quantities of fine-grained actinolite have developed within the magnetite-rich laminated zones of the cores from holes 5, 12, 18, 21, 22, 25, 26, 30, 32, 33, and 35; trace amounts can be found in almost all the cores. Minor quantities of fine-grained cummingtonite occur in the laminated zones of the cores from holes 28, 33, 34, and 35, and traces of it in most of the remaining cores, though none was observed in those from holes farther east than hole 20.

Medium- to coarse-grained andradite, occurring mainly within the quartzose layers, is the most distinctive mineral of unit G. Where this mineral has formed, the rock has been locally reconstituted into a lime-silicate granofels, and is locally associated with trace amounts of diopside, hedenbergite, calcite, actinolite, and epidote. Its occurrence and abundance have an unexpected lateral variation in that it is commonly more abundant in cores from holes relatively far from the gabbro, such as those from holes 25, 30, 32, 33, and 35. Minor quantities of the garnet occur in cores from holes 19, 20, 22, 27, 28, and 34; none were found in the cores from holes 18, 21, and 26. The moderate quantities of andradite present in cores from holes 4 to 6 perhaps owe their origin to the thermal effects of nearby pegmatitic veins.

Minor amounts of fine- to medium-grained hedenbergite, commonly accompanied by traces of interstitial calcite, are found in all cores except those from holes 4, 6, 20, 22, 26, 32, and 33. Coarse-grained plagioclase (andesine) and small tabular clusters of fine-grained biotite occur within a vein-like quartzose portion of a specimen from hole 20: the fine-grained feldspar and biotite associated with the cummingtonite in the adjacent laminated magnetite zones are all presumed to be, in part, of metasomatic origin. Metasomatic plagioclase was also found in core from hole 18.

In the core that probably corresponds with unit G from hole 27E, the dark layers consist largely of fine-grained stilpnomelane and finely disseminated magnetite in a fine-grained quartzose matrix. Very fine-grained green chlorite occurs as thin lenticular vein-like layers in the matrix, accompanied by trace amounts of carbonate and stilpnomelane. A few veinlets of stilpnomelane also follow the bedding of the core.

Submember F. Submember F is characterized by relatively thick, regularly-bedded zones of interlayered magnetite and cummingtonite lamellae separated by thinner quartzose layers that generally show a granule fabric. The high degree of preservation of the relic lamellar bedding affords many excellent examples of penecontemporaneous slumping, folding, and faulting of the laminated zones during sedimentation. Apparently many of these strata attained a sufficient degree of competence to be distorted without being completely destroyed, and all degrees of disruption from a slightly folded group of lamellae to a well-formed edgewise intraformational conglomerate can be found in many cores. These structures are interpreted to be primary rather than secondary because numerous "folded" and "brecciated" structures in many cores are truncated by undisturbed laminated strata above the disrupted zones.

Unit F is persistent and is readily recognized in most cores from the holes that extended downward to this unit, namely holes 4 to 6, 18 to 22, 25, 27, 30, and 32 to 34. Unit G, however, could not be clearly separated from unit F in cores from holes 19, 20, and 27E, where the combined thickness of both units is $40\frac{1}{2}$, 40, and $50\frac{1}{2}$ feet, respectively. The thickness of unit F ranges from 14 feet in hole 18 to $27\frac{1}{2}$ feet in holes 33 and 34 but averages approximately 15 feet in the eastern holes and about 24 feet in the western holes.

The distinctive bedded taconite-strata consist of closely spaced groups of interlayered lamellae, alternately rich in magnetite and in cummingtonite, which collectively form thick shaly bedded zones that range from about $\frac{1}{2}$ to 8 inches thick, although most are from 2 to 6 inches thick. The individual lamellae normally range from $\frac{1}{16}$ to $\frac{1}{8}$ inch thick and as a rule are regularly bedded and uniformly thick. Magnetite anhedra

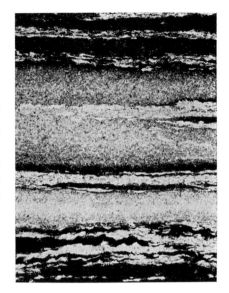

FIGURE 23. — *Shaly bedded taconite.* The intimate association of very fine- to fine-grained cummingtonite and magnetite in the *shaly bedded (cummingtonite-magnetite) cummingtonite-quartz taconite* of unit F can be seen in this photograph of the thinly laminated portions of these rocks. Relic lamellar structures of magnetite (black) are largely preserved but are considerably modified in detail by the recombination of quartz and magnetite into a decussate matrix of cummingtonite (light stippled gray). Minor amounts of fine-grained plagioclase, biotite, and traces of chalcopyrite are locally found with the amphibole in this and other sections. (25-102; 6X)

occurring as relic lamellae, as in Figure 23, are consistently fine-grained everywhere; even in cores from holes near the gabbro these grains seldom exceed ¼ mm in diameter.

Fine-grained acicular to prismatic cummingtonite is generally present within, and particularly between, the magnetite-rich lamellae in most cores. Beginning with core from hole 20, it becomes progressively more pronounced and locally abundant in cores from holes farther to the west. Typical grain fabrics occurring within the cummingtonite-magnetite-rich zones of the unit are shown in Figure 23. In this specimen, as in others, traces of plagioclase and brown biotite are intimately associated with the amphibole. Both cummingtonite and biotite occur within, and adjacent to, the lamellae and have partly replaced and slightly modified these relic structures during metasomatism. Some fine-grained actinolite, found in cores from holes 27 and 35, occurs similarly within the laminated zones.

A green-black variety of cummingtonite of unusual tabular habit, parallel to (100), occurs prominently in the laminated zones if only within a rather restricted area. Cummingtonite of this habit, locally abundant in the cores of holes 18, 19, and 22, and in moderate amount in core from hole 20, was not observed elsewhere. The petrographic description of its occurrence is essentially identical to that presented by Figure 24, which is discussed below, in the description of unit E.

The massive taconite-strata interbedded with the shaly bedded zones consist of quartzose layers ranging in thickness from ⅛ to 12 inches, although most are between ½ and 6 inches — averaging 3 inches — thick. The quartz of these layers is generally fine-grained, though the grains locally have diameters up to ½ mm in many of the cores from holes near the gabbro. Minor amounts of fine-grained cummingtonite are sparsely distributed in the quartzose layers. The quartzose layers generally show a granule fabric, slightly modified by recrystallization, but still clearly defined by disseminated fine-grained magnetite in the quartzose matrix. Numerous granules in cores from holes 27, 33, 35, and 27E also contain some hematite in addition to magnetite. In spite of the slight degree of recrystallization, at least some small part of each granule is relatively unaffected and contains "dust-sized" magnetite and hematite grains that seldom exceed a few microns in diameter. There is no obvious evidence as to the relative age of these dusty oxides within the relic granule structures. In core from hole 35, hematitic granules occur in the part of the rock containing a late, metasomatic, acidic feldspar-stilpnomelane-calcite veinlet. Similarly, within the core from hole 27E, hematitic granules occur principally where trace amounts of pale green minnesotaite, green and brown stilpnomelane, and ankerite have developed among the magnetite-quartz granules. These latter occurrences suggest that the hematite is martite, but the evidence is not conclusive.

Numerous small pods and irregular masses of coarse-grained andradite, rarely up to 2 inches thick, occur in core from hole 25; several minor, medium-grained occurrences were noted locally in the cores from holes 6, 21, 27, 30, 32, and 33. The occurrence of the garnet is essentially identical to that in unit G, being almost exclusively restricted to the quartzose beds of the unit. Minor quantities of medium-grained hedenbergite, accompanied by traces of calcite, are present in cores from holes 21 and 35, the latter of which also contains traces of medium-grained epidote. A few medium-grained anhedra of ferrohypersthene occur in the core from hole 5.

Submember E. Unit E consists largely of massive quartz taconite that contains numerous magnetite-bearing granule structures and distinctly fewer magnetite-rich laminated zones than the adjacent units. The presence of a few quartzose septaria structures supports the correlation of this unit with the septaria-bearing strata recognized by Grout and Broderick (1919, p. 17).

The submember was recognized in cores from all drill holes that extended to this horizon, namely holes 4 to 6, 18 to 22, 25, 27, 30, and 32 to 34. The unit has an average thickness of about 6 feet, but varies from $8\frac{1}{2}$ feet in hole 32 to $2\frac{1}{2}$ feet in hole 33. Although units D and F were identified in core from hole 27E near Mesaba, unit E was not apparent in this core.

This unit is composed largely of massive taconite-strata consisting of fine-grained quartz layers that show a distinct granule fabric. The quartzose beds generally range from $\frac{1}{4}$ to 6 inches thick, but most are from 2 to 4 inches. The numerous granule structures are defined by very fine-grained magnetite within quartzose layers; however, most of the magnetite in this lean submember occurs in the few laminated strata of the unit. The granule fabric is locally obscure in cores from the eastern holes as a result of recrystallization of the layers into medium-grained quartz. Septaria-like structures, consisting of whitish quartz-filled "fractures" within granule-bearing quartz layers, were found only in cores from holes 30 and 33. The quartz that constitutes the septaria structures is, in part, coarse-grained, with some grains several millimeters in diameter. In accordance with the sparse distribution of septaria structures found in outcrop, it is not surprising that septaria are rarely found in drill core.

In contrast to the adjacent units, laminated strata rich in fine-grained magnetite are fewer in number and rarely exceed 2 inches in thickness; most are less than ½ inch thick. Numerous granule-shaped clusters of fine-grained magnetite are commonly associated with these lamellae to collectively form relatively thicker, poorly bedded, magnetite-rich zones ranging from ½ to 6 inches thick, although most are 1 to 3 inches thick.

Fine-grained cummingtonite and some actinolite are almost always associated with the magnetite-rich zones, in some places forming layer-like zones up to ½ inch thick among these bedded taconite-strata. Cummingtonite, locally abundant in cores from holes 18, 19, 22, 25, 27, and 33, occurs in moderate quantities in the cores from the remaining holes. Moderate amounts of actinolite are found in cores from holes 18, 22, and 32, but it is present only to a minor extent in all other cores. Microscopic study reveals that relic granules and lamellae are still preserved in outline by disseminated fine-grained anhedra within quartz, cummingtonite, and (or) actinolite. Prismatic cummingtonite cuts all earlier structural features and commonly contains minute grains of radioactive zircon that have produced pleochroic haloes within the amphibole. Except for the presence of biotite, feldspar, and tabular cummingtonite, the microscopic appearance of these assemblages is similar to that shown in

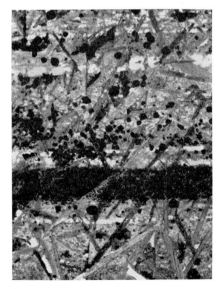

FIGURE 24. — Tabular cummingtonite. Even in highly reconstituted taconites, indications of initial lamellar structures are often evident, as delineated here by the densely packed zone of fine magnetite grains. The enclosing rock consists largely of a decussate mat of greenish, elongate, coarse-grained, tabular cummingtonite. In addition to numerous magnetite euhedra (black), the remaining matrix contains medium- to fine-grained prismatic cummingtonite, biotite (dark gray patches), and quartz. (19-20; 9X)

Figure 24. Where the magnetite-rich zones contain substantial amounts of actinolite, the mineral fabrics shown by its development in the adjacent granule structures closely resemble the result of metasomatic activity illustrated in Figure 21.

Medium- to coarse-grained euhedra of green-black cummingtonite, with an unusual (100) tabular habit, occur among some of the magnetite-rich lamellae in cores from holes 19 and 21. Even smaller grains of this habit might be present, though difficult to distinguish from the fine grains of normal prismatic cummingtonite that form most of the surrounding matrix. Relics of primary lamellae are still evident in these cores (Fig. 24). Cummingtonite of tabular habit is commonly accompanied by trace amounts of incipiently-twinned plagioclase (oligoclase?), brown biotite, and some chlorite grains. Numerous minute grains of zircon have produced pleochroic haloes within the enclosing biotite and both varieties of cummingtonite.

Coarse-grained anhedra of ferrohypersthene occur in core from hole 19, and several irregular masses of very coarse-grained porphyroblasts of hedenbergite are found in core from hole 18. Both pyroxenes have been partly replaced by cummingtonite and actinolite. Traces of andradite were noted in cores from holes 25, 30, and 32, and traces of epidote from holes 25 and 30. Small blebs of chalcopyrite occur in a calcite-quartz veinlet in core from hole 25.

Submember D. Unit D is characterized by the presence of lenticular layers to nearly pebble-shaped lenses of white chert-like quartz interbedded with zones containing numerous magnetite lamellae. These laminated zones of magnetite normally contain locally abundant silicates and as such are similar to those occurring in the overlying unit, but the latter unit is normally devoid of the small, highly lenticular quartzose layers.

Drill holes 2, 4 to 6, 18 to 21, 25, 27, 30, and 32 to 34 cut this stratigraphic interval. Unit D can be recognized in all but hole 5. The thickness of the submember ranges from 4 feet in hole 18 to 11 feet in hole 21, and a probable average thickness in the remainder of the holes is about 7 feet. The lithologic distinction between units C and D is less clear in core from hole 27E; the bedding features suggest a closer correlation with rock types occurring in unit D, which is the uppermost unit cored in this hole and is 21 feet or more thick at this point.

Numerous lamellae of fine-grained magnetite generally occur in closely spaced groups or laminated zones that range in thickness from $\frac{1}{4}$ to 3 inches, although most are about $\frac{1}{2}$ to 1 inch thick. Most of these magnetite-rich zones are uniformly thick or only slightly lenticular, while the bedding planes of the lamellae are commonly concordantly "draped over" or "bend around" the interbedded lenticular quartzose layers. Minor to moderate amounts of fine-grained actinolite, and generally lesser amounts of fine-grained cummingtonite, occur within the magnetite-rich bedded taconite-strata described above. Actinolite, however, is locally abundant in cores from holes 18, 25, 27, 30, 32, and 33, and cummingtonite occurs in moderate quantities in cores from holes 25, 27, and 33.

The massive taconite-strata of the unit generally consist of very fine- to medium-grained chert-like quartz layers that range in shape from thin beds of uniform thickness to small roundish pods or lenses less than an inch in diameter. The quartzose layers in the cores from the eastern holes are commonly medium-grained and distinctly lenticular, and range in thickness from ⅛ to 3 inches, but layers of from ¼ to 1 inch thick are most common. In hand specimen the quartzose layers are generally devoid of obvious granule structures though distorted remnants of these structures are commonly found in thin sections of both the magnetite- and quartzose-rich beds of the cores, as shown in Figure 25. The small

FIGURE 25. — *Laminated taconite.* The dark, magnetite-rich laminated beds and the light quartzose beds both contain probable relic granule structures of magnetite. Thin veinlets of magnetite and quartz, and minor amounts of fine-grained cummingtonite and actinolite occur throughout. (32-117; 9X)

roundish magnetite clusters with the vague tail-like appendages within the quartzose layers and among the magnetite lamellae have been interpreted as probably representing modified granule structures. These and the presence of several magnetite-quartz veinlets cutting the quartzose layers appear to indicate that some of the constituents of magnetite were locally mobilized during metamorphic reconstitution.

Moderate amounts of medium-grained hedenbergite, generally accompanied by traces of interstitial calcite, occur locally in the cores from holes 18, 21, 30, and hole 32, where it is also associated with minor quantities of fine-grained andradite. Anhedral grains of ferrohypersthene in core from hole 19 locally reach 1 cm in diameter.

Fine-grained stilpnomelane, minnesotaite, and ankerite accompany magnetite to constitute most of the dark-colored bedded taconite-strata in the core from hole 27E. As a result of recrystallization, these darker beds are not clearly laminated and the now subhedral magnetite is locally fine- to medium-grained, distinctly coarser than in most specimens from holes farther to the east. The interbedded fine-grained quartzose beds contain minor quantities of stilpnomelane and magnetite within poorly defined chert granules. Some parts of the cores are cut by quartz-stilpnomelane-ankerite veinlets, the constituents of which might conceivably have been mobilized from the adjacent ankerite- and stilpnomelane-rich beds by lateral secretion. On the basis of microscopic study and the relations seen in the field, however, it is the writers' impression that the constituents that resulted in the deposition of ankerite and stilpnomelane in the adjacent wall rock were probably introduced by solutions originating from the veinlets.

Submember C. Unit C generally consists of regularly-bedded laminated zones of silicates and magnetite intercalated with uniform quartzose layers, whereas the underlying unit normally contains distinctly lenticular beds of quartz. This is the uppermost unit of the Biwabik formation containing appreciable quantities of magnetite.

Core from the entire submember, or substantial parts of it, was recovered from holes 2, 5, 18, 21, 25, 30, and 32 to 34. The average thickness of the unit is about 42 feet but ranges from 36 feet in hole 5 to 47 feet in hole 25.

FIGURE 26. — Relic bedded chert. The dark areas represent relic bedded chert layers that contain abundant quantities of "micron-sized" magnetite grains; the whitish spots are small areas of incipient recrystallization into nearly pure quartz. The light gray parts of the rock have been completely reconstituted, largely into a decussate matrix of quartz and cummingtonite that has also partly replaced a few hedenbergite (H) prisms. (32-84; 9X)

The bedded taconite-strata of the unit typically consist of regularly bedded fine-grained magnetite lamellae which occur singly and also as closely-spaced groups that form uniform laminated zones up to $\frac{1}{2}$ inch thick. The magnetite lamellae can be found within quartzose material but most of the laminated magnetite zones are associated with a variety of silicates which also commonly extend outward from the laminated zones as much as $\frac{1}{4}$ inch. In many places the silicates that enveloped adjacent laminated magnetite zones meet, or nearly meet, to collectively form relatively thick (up to 10 inches) zones that now consist largely of silicates. Most of these magnetite- and silicate-rich zones, however, are generally about 3 to 6 inches thick and are most commonly found in cores from some of the holes near the gabbro, as discussed below.

The massive taconite-strata consist of locally silicate-bearing quartzose layers, mostly $\frac{1}{4}$ to 1 inch thick, although variations from $\frac{1}{16}$ to 5 inches can be found in most cores. Most of the quartzose material is very fine-grained and chert-like, except in cores from the eastern holes where it consists largely of fine- to medium-grained, and to a lesser extent of coarse-grained, quartz. Excellent examples of relic bedded to laminated chert layers can be seen with the microscope, as illustrated in Figure 26. In this example, the relic chert fabric is preserved by extremely finely divided magnetite, generally less than a few microns in diameter, within very fine-grained quartz. Some of the quartz has been slightly recrystallized locally during metasomatism, and in these places the primary chert texture is largely destroyed. Granule structures are uncommon within the quartzose layers. Near the bottom of the unit the quartzose layers become slightly lenticular, somewhat resembling those of the underlying unit D.

A variety of silicates are generally associated with the laminated magnetite zones, and to a lesser extent with the quartzose layers, of this unit. Minor to trace quantities of fine-grained fayalite occur as remnants or as inclusions within medium- to coarse-grained ferrohypersthene, which is found in moderate amounts in and surrounding the laminated magnetite zones of core from holes 5, 6, 18, and 21. Traces of fayalite also occur in core from hole 32. Some fine- to medium-grained poikilitic cummingtonite is locally found with ferrohypersthene in the laminated magnetite zones (Fig. 27). Medium-grained hedenbergite, locally abundant within the quartzose layers in cores from holes 5, 6, 18, and 21, is similarly found to a minor extent in cores from holes 25 and 30. Fine-grained cummingtonite is locally abundant in the quartzose and magnetite-rich parts of all the cores. All the anhydrous silicates can be found as remnants in, or have been partly replaced by, this amphibole. It is commonly associated with minor quantities of fine-grained actinolite and hornblende in cores from holes 25, 30, 32, and 33. Traces of pyrrhotite are locally found in the silicate-rich parts of the cores from holes 6, 18, and 21, accompanied by chalcopyrite in core from hole 6.

Several thin veinlets of hisingerite, accompanied by traces of fine-grained magnetite, cut the cores from holes 6, 30, and 32 in a direction essentially parallel to the bedding. A few cross-cutting veinlets of hisingerite cut the cores from holes 18 and 21. In most of these cores the earlier silicates have been incipiently to completely altered to a hisingerite- or nontronite-like material adjacent to the veins.

FIGURE 27. — Laminated taconite. Preservation of relic bedding structures is shown in this completely recrystallized laminated (silicate-magnetite) silicate-quartz taconite. The dark, magnetite-rich lamellae also contain moderate quantities of coarse-grained ferrohypersthene (dark gray) and traces of cummingtonite. The lighter, granoblastic quartzose zones contain a few small remnants of fayalite that have been largely replaced by minor amounts of cummingtonite, hornblende, and hornblende+cummingtonite (light grays). (21-128; 6X)

Submember B. This unit consists almost entirely of irregular layer-like masses of pyroxenes and amphiboles that have developed within vaguely bedded chert-like quartzose taconite. Consequently, although the unit as a whole is indistinctly bedded, it generally has a pronounced banded structure.

Submember B was cored in only five holes, 2, 18, 21, 25, 32; thickness of the unit in each is 15, 14, 17, 14, and 20 feet, respectively.

Normally a light green diopside is the most abundant pyroxene, though moderate amounts of darker green hedenbergite are generally also present in most cores. Both pyroxenes are mostly medium-grained granoblastic in texture, but many subhedral prisms of each exceed 5 mm in length in some places. These minerals occur mainly as irregular layer-like masses from about ¼ to 3 inches thick, mostly 1 to 2 inches on an average, and only traces are found in the adjacent quartzose layers. A portion of a metasomatized specimen from hole 21 consists of a medium-grained granoblastic mosaic of diopside, sphene, silicic feldspar (albite?), apatite, actinolite, chlorite and quartz; hedenbergite in the adjacent rock locally occurs as grains 15 mm in diameter.

In some cores the earlier pyroxenes have been partly to almost completely replaced by fine-grained cummingtonite and, to a lesser extent, by actinolite. The irregular patches of cummingtonite that replace the pyroxenes locally occur as thin vein-like layers that range in thickness from about ½ to 2 inches. Both the layer-like masses of cummingtonite and those of the unreplaced pyroxenes slightly resemble layers of sedimentary origin, but neither are relic sedimentary structures. A highly metasomatized specimen from hole 32 contains remnants of diopside within a rock now consisting largely of a decussate mat of coarse-grained hornblende (containing many pleochroic haloes and some secondary stilpnomelane), fine-grained cummingtonite and actinolite and traces of minute apatite prisms within a matrix of medium-grained granoblastic quartz.

Most of the interbedded layers range from 1/8 to 1 inch thick and consist of poorly to indistinctly laminated chert-like material. Although the relic sedimentary laminations are best preserved in the extremely finegrained quartzose layers, thin sections show that these structures are largely destroyed by even slight recrystallization of the layers into fine-grained quartz. On the other hand, relic layering and laminations are locally enhanced by the development of minor amounts of very fine-grained cummingtonite, actinolite, and traces of stilpnomelane (Fig. 28). The stilpnomelane of this specimen is the last of these metasomatic minerals to form and in many places is retrograde after actinolite.

Magnetite is generally uncommon; a few wispy or discontinuous fine-grained magnetite lamellae within the quartzose layers occur near the bottom in cores from most of the holes. Many fine-grained anhedra of pyrrhotite and chalcopyrite occur within the pyroxene- and amphibole-rich parts of the cores from holes 21 and 25, and chalcopyrite is similarly found in core from hole 18.

Submember A. The uppermost stratigraphic unit of the Biwabik iron-formation is a coarse-grained calcite marble, easily distinguished from the apparently conformable, overlying Virginia formation in the Eastern Mesabi district. According to what little information is available, the marble thins out and eventually vanishes westward from Mesaba. In contrast to the lime-silicate-rich unit B that underlies it, unit A rarely contains more than 10 to 15 per cent of lime-silicates.

FIGURE 28. — Stilpnomelane. Relic laminations (dark) in very fine-grained quartzose taconite consist of locally abundant aggregates of fine-grained stilpnomelane and traces of actinolite remnants. The light patches where the relic bedding structure has been destroyed represent zones of slight recrystallization now consisting of fine-grained quartz with minor amounts of metasomatic actinolite, cummingtonite, apatite, stilpnomelane, and some diopside elsewhere in the section. (25-39; 9X)

Only five holes, 2, 18, 21, 25, and 32, cut this horizon; the thickness of the unit in each of these cores is 2½, 4, 4½, 3½, and 7½ feet, respectively.

In core from hole 32 the marble ranges in color from nearly white to a light bluish-gray; farther east it is gray to gray-black. Under the microscope the only apparent difference between the darker and lighter colored varieties is in the relative abundance of unidentified minute black specks (graphite?) occurring as inclusions within the calcite grains. Most of the silicates occurring in all the cores are too fine-grained to be readily recognized in hand specimen, but they are commonly abundant enough to locally give the marble a greenish tint. Thin sections of these silicate-rich marbles reveal that fine-grained anhedra of diopside, traces of wollastonite and, rarely, quartz, occur interstitially to the granoblastic mosaic of calcite grains. Very coarse-grained prisms of wollastonite are locally abundant along with coarse-grained diopside in core from hole 32. Although this assemblage might possibly be the result of thermal metamorphic effects of the Duluth gabbro or a nearby diabase sill, the mineral assemblages in unit B of the core from this hole indicate that it is the result of its proximity to a pegmatitic vein. Minor quantities of medium- to coarse-grained andradite have developed in cores from holes 18, 21, and 32. Minor numbers of medium-grained idocrase anhedra occur interstitially to the calcite grains in the marble core from hole 21. This core has been subsequently cut by thin veinlets of greenish, wax-like nontronite, which also extend into the intergranular regions within the granoblastic mosaic of calcite grains. Some parts of the veinlets are distinctly reddish, presumably due to the presence of finely disseminated ferric oxide. Small anhedral blebs of chalcopyrite and pyrrhotite occur in trace amounts within the core from hole 25.

VIRGINIA FORMATION

In the Eastern Mesabi district, the area underlain by the Virginia formation consists of a swamp-like depression. There are no outcrops of the formation near the Mitchell pit, but some occur a few miles to the southwest. Some specimens of the formation were obtained as core from holes 2, 3, 18, 21, and 25.

In this district most of the Virginia formation can be described as a somewhat laminated, fine-grained, light gray quartzose hornfels, locally rich in biotite. The formation is generally described as an argillaceous rock on the Main Mesabi range, but rocks of this nature were not found in the Eastern Mesabi district. In many places a binocular microscope is required to distinguish the metamorphosed formation from some of the fine-grained varieties of metamorphosed diabase sills that locally occur within it. Additional specimens of the Virginia formation became available after the present brief study of the formation had been completed. Consequently, a detailed description of the metamorphism of the formation will appear elsewhere at a later date.

The contact metamorphic effect of the gabbro upon the Virginia formation, as noted by Grout and Broderick (1919, p. 9), is a very complete recrystallization into a sugary textured hornfels, which extends in some places several hundred feet. Because the specimens of the formation collected for the present study were obtained from holes both near and far from the present gabbro contact, their descriptions will serve to briefly outline some of the results of thermal metamorphism induced by the gabbro.

In cores from holes 2 and 3, within a few hundred feet horizontally of the present gabbro contact, the formation has been completely recrystallized to a medium- to fine-grained quartz-biotite-anorthoclase (orthoclase?) hornfels. Although the concentration of biotite into lamellae probably reflects initial lamination, the lamellae are not foliated but characteristically have a decussate fabric that indicates final recrystallization probably took place mainly under thermal rather than stress conditions. Minute blebs of interstitial pyrrhotite occur locally. In addition, core from hole 2 contains trace amounts of wollastonite, diopside, and iron-rich cordierite, most of which has altered to a very pale green chlorite-like material resembling prasiolite.

In core from hole 21, about 4,000 feet horizontally from the gabbro, recrystallization has yielded a fine-grained chlorite-white mica (muscovite?)-quartz-orthoclase hornfels that contains biotite only in trace amounts. These minerals are generally more "murky" and clouded by inclusions than those occurring in the specimens nearer the gabbro. A thin, impure, carbonate-rich zone within the Virginia formation from this core has recrystallized to produce a complex lime-silicate calcite marble consisting mainly of an interlocked mosaic of calcite grains that are "dusty" with many minute unidentified inclusions. These inclusions are presumed to be mineralogically the same as the larger grains of diopside, idocrase, sphene, periclase, and leucoxene that occur interstitially.

Farthest away, in core from hole 25, almost two miles from the present gabbro contact, it is evident in thin section that some parts of the formation were originally immature quartz graywackes prior to recrystallization. The angular or "gritty" fabric of the quartz grains and a few altered (sericitic?) fragments of feldspar are still apparent. The surrounding fine-grained matrix material is typically cloudy and consists mainly of slightly recrystallized grains of quartz, white mica (muscovite or sericite), and some biotite, chlorite, and orthoclase.

DIABASE DIKES AND SILLS

A diabase dike, 35 feet thick, occurs in the Upper Cherty member near 6,600 feet west along the baseline from the eastern end of the Mitchell pit and another dike is partly exposed in a small drainage ditch east of the pit. Diabase sills, ranging from 1 to 15 feet thick, are present in core from holes 5, 6, 7, 18, 21, and 32. A diabase sill, about 6 feet thick, is also exposed along the southern edge of the Erie Mining Company pit northeast of Mesaba. These sills characteristically occur in the upper beds of the Upper Slaty member of the iron formation and near the bottom of the Virginia formation. Recent drilling east of Mesaba by the Erie Mining Company has recovered core from diabase sills within the Upper Cherty member.

The diabases in general are dark gray and fine-grained, and in some places are difficult to distinguish in hand specimen from the metamorphosed Virginia formation, particularly where the latter is hornfelsic. In hand specimen, the rock has a barely discernible diabasic texture consisting of small laths of plagioclase (labradorite) with interstitial fine-grained anhedra of augite or other secondary mafic minerals.

Although the diabase bodies do not appear metamorphosed in hand specimen, the effects of metamorphism are obvious in thin section. Within the least affected diabases, the edges and corners of the plagioclase laths are incipiently corroded and resorbed and the interstitial grains of augite are slightly recrystallized. Characteristic of the effects of moderate contact (thermal) metamorphism, this fabric is common in the Eastern Mesabi district. Evidence of this nature might possibly have been the basis of the statement by Grout and Broderick (1919, p. 8) that their study of the dikes and sills indicated these rocks had been metamorphosed by the contact action of the gabbro and therefore can not be considered as apophyses extending from the gabbro. In numerous other cases, however, the original augite of the diabase has been replaced by yellowish to greenish amphiboles and to a lesser extent by pale brown biotite. In these rocks the diabasic fabric of the rock is preserved, but a lattice-like network of rod-like grains of ilmenite is superimposed on the rock texture. In these specimens the metamorphism was necessarily accompanied by the addition of water in order to form these new hydrous minerals.

No evidence points to metamorphism of the iron formation by the diabase intrusions, perhaps because both rocks have been subsequently highly metamorphosed. The formation of wollastonite, andradite, and diopside in core of unit A near to a diabase sill in hole 32 might be related to the sill, though it has been interpreted to be a part of the thermal effects upon the marble of unit A by a nearby pegmatic vein. There is clear evidence, however, that the sills were highly altered in places where the adjacent iron formation was partially metasomatized. In these cases an alteration zone up to 1 inch thick and paralleling the contact is present within the diabase. In hand specimen this altered zone looks like a chilled margin of the sill; in thin section it is characterized by an advanced state of obliteration of the diabase fabric. There the feldspar laths are commonly destroyed or altered to "white mica" (sericite?) and the mafic minerals replaced by a matrix of extremely fine-grained minerals that cannot be resolved with the microscope. Within these altered zones, though some relics of the diabase minerals persist, they are locally associated with veinlets and scattered grains of actinolite and diopside of metasomatic origin.

DULUTH GABBRO

No fresh exposures of the Duluth gabbro are found near the Mitchell Pit. Core from holes 1 and a number of shallow holes from the crusher site, a few hundred yards south of hole 1, nevertheless provided many gabbro specimens. Fresh artificial exposures of the gabbro are also available in the ore conveyor tunnel at the base of the 60-inch diameter crusher. Though contacts between the Duluth gabbro and the Virginia and Biwabik formations are concealed, the coarsely recrystallized parts of the iron formation exposed at the eastern end of the Mitchell pit probably lie within 500 feet horizontally of the gabbro.

In hand specimen this rock is a light to dark gray, medium- to coarse-grained gabbro that shows a variety of textures ranging mainly from hypautomorphic granular (with a diabasic to a trachytoid fabric) to xenomorphic granular and locally shows a pseudo-layered or stratiform ("banded") structure. Fine-grained anhedra of olivine and augite occur interstitially to subhedral laths of plagioclase. Minor amounts of chalcopyrite and pyrrhotite are locally abundant in a few core specimens, and in some places these sulfides have completely replaced the olivine and augite while still preserving the diabasic fabric of the plagioclase laths. The sulfides are characteristically associated with the mafic minerals of the rock, particularly coarse-grained brown biotite. It is not certain whether the sulfides and biotite were deuteric or post-magmatic (hydrothermal) in origin, but probably both modes of origin were effective locally. The plagioclase laths of some specimens have been altered to a chalky white, clay-like material, which locally accentuated the diabasic fabric of the rock, nearly always accompanied by abundant, coarse-grained brown biotite.

In thin section most of the Duluth gabbro in this region can be classified as olivine gabbro, but some parts of the complex are either simple (normal) gabbros or troctolites. The plagioclase grains in all these varieties are labradorite (ca. An_{70}) laths which possess incipiently developed spindle-shaped pericline twins that somewhat complicate the dominant albite twinning. A few grains are faintly zoned (normal) with slightly less calcic rims, although this feature is not common. A few anhedra of quartz occur interstitially to plagioclase but are rare.

The olivine in the troctolites and in olivine gabbros normally occurs as anhedra interstitial to the plagioclase laths and most commonly forms a hypautomorphic granular rock with a coarse-grained diabasic fabric. The olivine grains of the olivine gabbros often have thin coronas of violet-tinted augite (Ti-rich?), which is most probably due to incomplete reaction between olivine and the melt in which it had formed, although it is conceivable that they might be the result of deuteric solutions. The augite in the normal and olivine gabbros also generally occurs interstitially to plagioclase in a fashion similar to olivine, although in some places it poikilitically encloses olivine and less commonly plagioclase to produce an ophitic fabric.

Brown biotite appears next in the sequence of crystallization, but it is not clear whether it was a late-stage deuteric mineral or a product of a later high-temperature hydrothermal alteration. In many places fine-grained biotite commonly surrounds or rims the mafic minerals or the minor amounts of accessory magnetite and it appears as if it might have been a late magmatic (deuteric) mineral. In many other places, however, it is coarse-grained, and intimately associated with highly altered feldspars, abundant prisms of apatite, replacement sulfides, and partly recrystallized parts of the gabbro, all of which might suggest later high-temperature alteration. In these occurrences faint pleochroic haloes or rims have locally formed in the biotite around poikilitically enclosed apatite crystals, presumably due to trace amounts of radioactive rare earths (Y, Ce?) possibly present in the apatite.

Some parts of the gabbro complex in this district were clearly altered by later solutions, probably aqueous in nature. In these places the gabbro contains small irregular masses and thin veinlets of a dark, waxy or serpentine-like material. On the basis of x-ray powder studies, most of this vein material is believed to be nontronite. Immediately adjacent to these veinlets olivine is invariably altered to a brownish nontronite or hisingerite-like mineral, or to a lesser extent, to a pale greenish mineral resembling bowlingite. In a few veinlets some "cross-fiber" zones of grains resembling chrysotile have formed, but x-ray studies indicate that it is a chlorite-like mineral with a basal spacing of 14A. Minor amounts of pyrite occur locally in the nontronite veinlets. With the exception of the sulfides and magnetite, nearly all the remaining gabbro minerals are slightly altered where next to those veinlets.

PEGMATITIC VEINS

From the detailed examination of the lateral variations of the mineral assemblages within the numerous stratigraphic units of the Biwabik iron-formation, it was found that anhydrous silicates, particularly medium- to fine-grained fayalite and ferrohypersthene, occur within broad zones that roughly parallel the present gabbro contact. The distribution and lateral variation of abundance of these minerals probably constitute the effects of the early stages of thermal metamorphism that accompanied the emplacement of the Duluth gabbro. In other parts of the formation where these silicates did not develop, the previously existing taconite, consisting largely of quartz and magnetite, was merely slightly recrystallized. There is not a single occurrence, or any evidence whatsoever, which indicates that the fayalite or ferrohypersthene had developed from a previous silicate or carbonate, nor do any relics of earlier silicates or carbonates exist in the rocks that contain no metamorphic silicates. During a later stage of the over-all metamorphic event, the development of medium- to coarse-grained hedenbergite, ferrohypersthene, hornblende, and medium- to fine-grained actinolite and cummingtonite was superimposed on the rocks described above. From hundreds of thin sections, innumerable core specimens, and extensive exposure of rock in the Mitchell pit, it is clear that the fayalite-bearing taconites and those consisting essentially of quartz and magnetite have been locally replaced by pyroxenes and amphiboles. The lateral distribution, abundance and grain size of the later minerals, particularly the pyroxenes, are quite sporadic and during the first stages of the study were difficult to explain. It was not until a detailed study of the miles of new rock exposures of the Mitchell pit could be made that these and other minerals were seen to be of metasomatic origin. The introduction of nearly vertical pegmatite-like veins, ranging from $\frac{1}{8}$ inch to 2 feet in thickness, into the Biwabik iron-formation during a later stage in the over-all metamorphic event was accompanied by locally extensive metasomatic activity in the adjacent wallrock. In these places portions of the existing metamorphic minerals and parts of still earlier magnetite-quartz assemblages were locally replaced by, or reconstituted into, calcium- and magnesium-bearing ferrous pyroxenes and amphiboles.

The composition of the veins can, and commonly does, vary abruptly within a few dozen feet, laterally or vertically, in the vein. At one extreme the veins are acidic or granitic, consisting mainly of quartz and pinkish alkali feldspar. The pinkish alkali feldspar, called orthoclase in the field, was found to range from albite to albiclase in composition. Spectrographic studies also indicate minor amounts of barium and strontium. At the other extreme the veins are mafic, resembling lamprophyres consisting mainly of hornblende. The veins commonly contain grains of feldspar and amphibole from $\frac{1}{2}$ to 2 inches in diameter although an average grain size would be nearer to $\frac{1}{4}$ inch: exceptional grains reach 6 to 8 inches. In the pegmatites these minerals are commonly accompanied by calcite, oligoclase-andesine, green biotite (now largely chlorite), and to a lesser extent by pyrite, apatite, magnetite, zircon, ferrohypersthene, andradite, muscovite, chalcopyrite, molybdenite, and loellingite. Minor amounts of retrograde stilpnomelane after hornblende, as well as interstitial stilpnomelane, have formed in almost all the larger mafic veins. The variety and extent of mineralization in the adjacent iron formation is directly dependent upon the mineralogy and thickness of the associated veins.

Wherever the veins contain hornblende, a variable zone of grains of this amphibole typically occurs in the immediate wallrock. Where the iron formation is quartzose, this amphibole-rich zone ranges from about $\frac{1}{8}$ inch thick next to the more acidic veins, and up to about 3 or 4 inches next to the most mafic veins; in some of these latter places the hornblende-forming solutions have permeated hundreds of feet along the bedding planes of the adjacent magnetite-quartz iron formation.

The most striking results of calcium metasomatism are seen in the development of abundant grains of hedenbergite, ranging up to one inch in diameter, in the taconite adjacent to pegmatites that are rich in hornblende. At one place a vein two feet thick consisting of mainly hornblende was bounded on each side by a massive zone of equal thickness of very coarse-grained hedenbergite. The abundance of this pyroxene then diminishes with distance from the vein to form layer-like masses up to 14 inches thick within about 25 feet of the vein. Up to about 200 feet from the vein, layer-like masses of the medium-grained hedenbergite can be found up to one inch thick, and numerous small porphyroblastic clusters, up to $\frac{1}{2}$ inch in diameter, occur as far as 300 feet from the vein. Commonly minor to trace amounts of calcite, chalcopyrite, apatite, and loellingite occur interstitially to, or partly replace, the hedenbergite in the wallrock where the pegmatites are locally rich in these minerals. Pyrrhotite commonly occurs in the wallrock where the veins contain pyrite. Field relations show that the hedenbergite-bearing assemblages occur in taconite consisting largely of magnetite and quartz. Petrographic evidence, particularly that showing partial replacement of relic sedimentary textures and minerals, indicates that these assemblages were formed metasomatically, directly from quartz and magnetite and locally from the previously existing metamorphic silicates, presumably by hot calcium- and magnesium-bearing aqueous fluids. Where the veins are particularly thick or rich in hornblende, the thermal effects were apparently sufficient to prevent the formation of amphibole within the inner hedenbergite-bearing zone and the intermediate recrystallized magnetite-quartz zone, although cummingtonite-bearing magnetite-quartz rocks are invariably found within an outer zone at some distance from the vein. Cummingtonite is here defined to be a calcium-poor, magnesium-bearing, ferrous amphibole containing from 50 to 80 Mol per cent of the iron end-member grunerite of the cummingtonite family, as discussed in Chapter 3.

Where the pegmatites are less massive and somewhat less mafic, the mineral zones about the veins are normally compressed and the outer amphibole-bearing zones have encroached upon the inner pyroxene-bearing zones. In every thin section examined, the amphiboles without exception are retrograde from, or clearly replace, the pyroxene or magnetite-quartz assemblages. It is important to note that small amounts of hornblende, actinolite, chalcopyrite, plagioclase, green and brown biotite, apatite, and zircon are commonly associated with the cummingtonite in many places. Pleochroic haloes around minute zircon grains occur in the amphiboles and biotites from both the pegmatites and the adjacent iron formation.

Some pegmatite veins are locally rich in albite and green biotite, in addition to containing hornblende. In a few of these places the adjacent iron formation has been strikingly metasomatized into a coarse-grained, quartz-magnetite-albite-hornblende taconite that is gneissic in appearance. Similarly metasomatized zones up to three inches thick, that are almost sill-like in appearance, commonly extend two to three hundred feet into the surrounding iron formation adjacent to the albite-rich veins. In a few places thin tabular zones up to two inches thick, containing only green biotite, cummingtonite, and zircon, are found extending from the veins several hundred feet into the iron formation.

At many places it appears that the introduction of calcium and water was distinctly later than the introduction of magnesium. In these places the development of clusters of very coarse-grained ferrohypersthene in the wallrock adjacent to the pegmatites was followed by partial replacement by actinolite near to the veins and by cummingtonite farther from the veins. It is not possible to establish from thin section examination whether these ferrohypersthene remnants developed from a previous existing silicate or not, because all earlier fabrics have been destroyed. Neverthless, because of their symmetrical distribution adjacent to the pegmatitic veins, it is believed they owe their origin to the metasomatic reconstitution of the iron formation immediately adjacent to the veins.

Where hornblende is not abundant in the quartz-feldspar veins, the adjacent iron formation has sparsely distributed metasomatic minerals. Though all the minerals mentioned above can be found in small amounts, the most obvious mineralization consists of small feldspathic patches that diminish in size and abundance into the adjacent taconite away from the veins. Few feldspathic patches are found beyond 100 feet of such pegmatitic veins.

At numerous places the pegmatites become abruptly quartzose, with the result that little if any metasomatic activity is apparent in the taconites adjacent to these veins.

3. MINERALOGY

INTRODUCTION

This chapter is essentially a generalized summary of the minerals and their typical modes of occurrence in the metamorphosed iron formation of the Eastern Mesabi district. The mineralogy of the iron formation of the Main Mesabi range farther west is discussed by Gruner (1946) and White (1954) and will not be repeated here.

The necessary optical properties required for the identification of some of the metamorphic ferrous silicates and chemical analyses of several important minerals found in the area are also given. All mineral samples were carefully hand-sorted, then crushed and screened. Preliminary concentration of each sample by means of a Franz isodynamic separator produced a product approximately 98 to 99 per cent pure. The samples were cleaned further by hand-picking while viewed with a binocular microscope. Each sample submitted for analysis was approximately 99.5 per cent pure. The chemical analyses of these minerals are given in Table 4. All mineral analyses were made by the Rock Analysis Laboratory, Department of Geology, University of Minnesota, Dr. Samuel S. Goldich, Director, with funds provided by the Minnesota Geological Survey.

TABLE 4. MINERAL ANALYSES OF SAMPLES FROM THE METAMORPHOSED IRON FORMATION OF THE EASTERN MESABI DISTRICT

(C. Oliver Ingamells, Eileen H. Oslund, and Doris Thaemlitz, analysts.)

	Fayalite	Ferrohy-persthene	Cummingtonite	Hornblende	Hedenbergite	Hisingerite	Calcite
SiO_2	33.58	50.07	54.67	45.73	49.56	38.19	.07
Al_2O_3	.18	.19	.27	5.21	.14	0.00	.10
TiO_2	.00	.00	.03	.03	.03	0.01	
Fe_2O_3	.46	.40	.73	3.70	1.31	19.91	
FeO	60.40	37.81	27.16	24.43	22.49	24.64	.87
MnO	1.48	.47	.80	1.38	2.13	0.66	1.90
MgO	3.94	10.09	13.66	5.99	3.63	2.36	.09
CaO	.18	.89	.99	9.94	20.20	0.61	53.24
Na_2O	.01	.05	.05	1.13	.24		
K_2O	.01	.04	.06	.56	.00		
H_2O+	.04	.17	1.27	1.46	.24	8.40	
H_2O-	.00		.00	.00	.09	5.53	.08
P_2O_5		.05					
CO_2							43.88
F				.28	.01		
Cl				.27	n.d.		
Total	100.28	100.23	99.69	100.11	100.07	100.31	100.23

Almost all the minerals are sufficiently abundant and coarse-grained for recognition with the hand lens or under the microscope. With the exception of some minerals that occur in accessory amounts, all have been x-rayed. In general, after a given mineral was identified by optical properties and then verified by x-ray methods, identification of other minerals was based upon similarity of optical properties and of appearance under the hand lens. This is the most reasonable method of identification since every mineral occurrence can scarcely be checked by x-rays for positive identification.

The discussion below is subdivided into the following sections determined by the mineral families: elements; sulfides; arsenides; oxides; carbonates; sulfates; phosphates; silicates.

ELEMENTS

Carbon (Graphite?). Finely divided soft black carbonaceous material with a black streak has been tentatively identified as graphite. It is invariably associated with, and virtually inseparable from, fine-grained quartz which precludes verification of its identity by x-ray methods. In the area studied, it was observed in recognizable amounts only within submember Q (the Intermediate Slate). It is possible that minor amounts of carbonaceous material (now graphite) are distributed throughout the formation.

SULFIDES

Sulfides are not abundant in the iron formation, but are found in minor amounts in many of the submembers throughout the area.

Covellite. Secondary after chalcopyrite, covellite occurs in minute amounts within a thin chalcopyrite- and pyrite-bearing pegmatitic veinlet found in the Mitchell pit, but it is generally rare.

Chalcopyrite. Locally found within the pegmatitic veins and in the adjacent iron formation, chalcopyrite is associated with the hornblende, quartz, and calcite of the pegmatitic veins and is normally subordinate to the pyrite that occurs with it.

Chalcopyrite is also found to a very minor extent in several submembers of the iron formation. It is not subject to any stratigraphic restrictions but typically occurs within local areas that have been metasomatized. Here it is present as small blebs, mainly within hedenbergite, actinolite, and cummingtonite. It appears to be more abundant, but still minor in extents, in core from hole 27E. Thin (1 cm) chalcopyrite-bearing zones have developed in the wallrock adjacent to several of the pyrite-bearing quartz and carbonate veinlets in the core from hole 27E. Chalcopyrite occurs disseminated in the gabbro of the area as previously noted.

Molybdenite. Small micaceous plates of molybdenite were found only in one thin hornblende-feldspar-quartz pegmatitic veinlet, observed in the Mitchell pit.

Pyrite. The most common of the sulfides found in the pegmatitic veinlets, pyrite has not been identified in the iron formation on the Reserve Mining property, although trace amounts might be associated with the somewhat commonly occurring pyrrhotite. Thin pyrite-bearing zones, usually accompanied by minor amounts of chalcopyrite, have developed in the wallrock near several of the pyrite-quartz-carbonate veinlets in the core from hole 27E.

Small pyrite cubes are locally abundant within many thin nontronite (after hisingerite) veinlets found within the iron formation in the area studied, but pyrite has not been observed in the fresh, unaltered hisingerite veinlets.

Pyrrhotite. The most abundant sulfide in the iron formation in the Eastern Mesabi district is pyrrhotite. Almost without exception it occurs where relatively coarse-grained silicates have formed, especially hedenbergite, ferrohypersthene, and fayalite. In a few of these places very minor amounts of chalcopyrite occur with the pyrrhotite. Although found in the Upper Cherty member, pyrrhotite is perhaps more common in the Upper Slaty, Lower Cherty, and Lower Slaty members, and appears to be most abundant locally in submember Q (the Intermediate Slate). On the presence of pyrrhotite in the gabbro, see pp. 71–72.

ARSENIDES

Loellingite. Minute amounts of silver-white loellingite have been found in the Upper Cherty member in the Peter Mitchell Pit. It occurs in highly metasomatized quartzose taconite containing abundant hedenbergite and hornblende. The arsenide occurs essentially interstitially to, or is intimately intergrown with, the metasomatic silicates. One specimen of a thin pegmatitic veinlet has been found which contains a few elongate, striated crystals of this mineral.

In polished section it is only faintly anisotropic and is medium hard, properties not normally associated with loellingite.

Its association with the metasomatic calcium-rich ferrous silicates suggests that it is of metasomatic origin.

OXIDES

Hematite. Constituting probably less than one tenth of one per cent of the iron oxides of the iron formation of the Eastern Mesabi district, hematite occurs mostly within the lower part of the Upper Slaty and the upper part of the Upper Cherty members in the holes south and west of Iron Lake. In these places much of the hematite is martite, pseudomorphic after small magnetite octahedra and anhedra. Many of the occurrences, however, consist of extremely fine-grained, disseminated hematite "dust" (with or without magnetite) in quartzose granules. Some quartzose granules have a light cream to almost brownish tan color due to extremely fine, disseminated opaque material, which the writers think might be various admixtures of clay-like material and hematite(?) dust.

MINERALOGY

In hole 27E, finely disseminated hematite was found in the form of granules, and Figure 29 is somewhat typical of their occurrence. The following description refers to the bottom of that photograph: very fine-grained magnetite is disseminated within quartzose granules and to a slightly lesser extent in the surrounding quartzose matrix. There are marked differences in magnetite content of the granules, some of which are magnetite-rich. This portion of the specimen contains the least altered or reconstituted relic structures. It is believed that carbonate-bearing fluids entered the system, resulting in the deposition of a mass of relatively coarse-grained calcite. Traces of minnesotaite have formed along

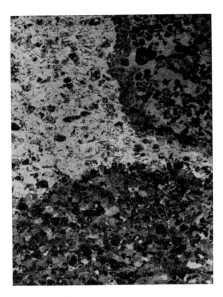

FIGURE 29. — Relic granule structures in calcite. A variety of relic granule structures occurs in this specimen. From a study of the cores and microscopic fabrics, it is believed that initially the entire specimen was essentially identical to that at the bottom of the picture. Here the rock consists of extremely fine-grained magnetite disseminated in quartzose granules that were little affected by recrystallization. Carbonate-forming solutions deposited coarse-grained calcite (upper right), in which delicate granule structures of dusty martite are now preserved. Slight recrystallization of the original rock into fine-grained quartz and magnetite (upper left) still preserved clear relic granule structures. (27E-99½; 3X)

the carbonate replacement front. Within this calcite mass, most of the fine-grained ferric oxide granules still perfectly preserve the internal dust-like dissemination of iron oxide displayed by the magnetite granules in the bottom part of the specimen. Although slight recrystallization of some of the granules has taken place, the process for this reconstitution might have been addition of oxygen or leaching of ferrous iron from the magnetite structure to account for the perfect preservation of the delicate granule fabric or structure of the ferric oxide granules. In this area, however, the matrix consists of calcite, with little if any interstitial disseminated iron oxides. The fluids that deposited the calcite apparently moved through the whole upper part of the specimen with the result that the magnetite-quartz granules have slightly recrystallized (magnetite now occurring as small octahedra and anhedra within and around the former granule outline) and are in a matrix of clear quartz that is relatively coarser than elsewhere in the specimen.

Magnetite. Next to quartz, magnetite is the most abundant mineral in the metamorphosed Biwabik iron-formation. Most of the magnetite is concentrated in thin layer-like aggregates representing primary sedimentary beds. Quartzose pebbles constituting intraformational conglomerates are also frequently magnetite rich. The most striking occurrence of magnetite is within the magnetite-quartz granules. These structures consist of an ovoid concentration of very fine-grained magnetite (grain size ranging from less than 1 micron to 20 or 30 microns in diameter) within fine-grained quartz, as illustrated in Figure 29. These granules consist essentially of magnetite "dust" finely disseminated within a primarily quartzose granule. Many granules typically show syneresis-like cracks, suggesting that the granules were once roundish pods of gelatinous material that became desiccated during or slightly after deposition. In many places these granule structures and their surrounding quartzose matrix have been recrystallized, as seen in Figure 29. The dusty appearance of the magnetite has been partly or almost wholly destroyed, but the granule-structure outline is commonly still preserved by a rim of anhedral grains to euhedral octahedra of magnetite (or in a very few places to martite) that range from about .01 to .25 mm in diameter but rarely larger; most of these grains lie between .02 and .1 mm in diameter. The magnetite-rich lamellae, layers, pebbles, and granules found in the area studied are interpreted to be relic sedimentary structures; structures of this type are found in every submember of the iron formation except the uppermost carbonate unit. J. W. Gruner (personal communication) has suggested that some of the black opaque "dust" within some of the granules may be, in part, graphite. Graphite has been tentatively identified within submember Q, the Intermediate Slate (Fig. 13). In this specimen, graphite could not be incorporated into the ferrohypersthene metacrysts during growth, and it appears to have been essentially pushed out of the space into which the silicate grew. Fine-grained magnetite granules from nearly every other submember have been partly replaced by the various ferrous silicates without the development of a similar or related occurrence of graphite. A possible exception may occur in submember P where granule-like outlines, defined by minute black opaque particles, are perserved in some fayalite metacrysts. Although it has been called magnetite, because the fragments are magnetic, some of the material may be graphite. The presence of graphite would not be unusual if one considers the nature of the original iron formation, but it seems probable that this minute black opaque magnetic material is magnetite. The fact that some parts of a given specimen may consist of martite preserving the identical fabric also seems to indicate that the original material was essentially all magnetite.

The grains of magnetite that occur in the lamellae, thin beds, and pebbles of intraformational conglomerates are usually anhedra that generally range from about .01 mm to .25 mm in diameter, but the majority of the grains have diameters of about .05 to 0.1 mm. Recrystallization of magnetite into coarser and more euhedral grains is quite sporadic, and only in a very general way does the grain size of magnetite reliably indicate proximity to the gabbro, except locally within about a quarter of a mile, horizontally, of the gabbro contact. In this area, grains of magnetite within many of the submembers commonly have diameters ¼ to ½ mm and some grains ranging up to 4 mm in diameter are found within a few hundred yards of the contact.

CARBONATES

Calcite. An important constituent of many of the pegmatitic veins, calcite is one of the last minerals to form in them. Within the pegmatites, calcite is associated with hornblende, quartz, pyrite, and some chalcopyrite. Several veinlets scattered through the iron formation consist of quartz and calcite, locally containing minor amounts of pyrite, but from the relatively few occurrences their genetic relationship to the pegmatitic veins is not clear.

The chemical analysis of a sample of calcite from a pegmatite occurring in the Mitchell pit is given in Table 4. Spectrographic study indicates trace amounts of strontium and barium in this specimen. No carbonates with indices approaching siderite were found in the Eastern Mesabi district. Slightly brownish-stained carbonates from veins in hole 27E were called siderite during logging of the core, but later index determinations, differential thermal analysis, and x-ray diffraction studies of some of these samples showed that many were calcite and ankerite as well as siderite. Spectrographic comparison of a calcite vein in submember V of hole 27E and the analyzed calcite showed their composition to be very similar: i.e., typically iron-, manganese-, and magnesium-poor, and titanium- and strontium-bearing.

Minor amounts of interestitial calcite, identified by differential thermal analysis and x-ray diffraction, are typically associated with hedenbergite that has developed within the taconite adjacent to some pegmatitic veins. In the western part of the area, it is also commonly associated with hornblende, actinolite, hornblende+cummingtonite, and rarely cummingtonite. Dolomite is genetically related with fine- to medium-grained cummingtonite in the Erie Mining Company pit No. 3.

Calcite constitutes almost all of submember A and prior to metamorphism probably constituted much of submember B. Arenaceous and argillaceous impurities in the primary limestones of these units probably were responsible for the formation of the lime-silicates during contact (thermal) metamorphism by the gabbro.

SULFATES

Barite. Barite was found only in submember S of hole 21 where it occurs in trace amounts in a granoblastic mosaic with calcite and quartz. Both the calcite and barite have been partly replaced by stilpnomelane, which also occurs interstitially to, and has slightly replaced, the quartz grains. Barite was introduced with the calcite.

PHOSPHATES

Apatite. A typical accessory mineral in the pegmatitic veins, small amounts of apatite are found in some places where the adjacent wallrock has been metasomatized. It is most commonly associated with hornblende, hedenbergite, hornblende+cummingtonite, plagioclase, and biotite. Apatite is the only phosphate recognized in the iron formation, and on the basis of its association and occurrence, is interpreted to be metasomatic, rather than a recrystallization product of initial constituents.

SILICATES

The metamorphosed Biwabik iron-formation on the Eastern Mesabi range is characterized by the abundance and the variety of ferrous silicates. The minerals in approximate order of abundance in the formation are cummingtonite (locally up to 60 per cent), fayalite, hedenbergite, the calcium-rich amphiboles, ferrohypersthene, andradite, and biotite (locally up to 5 per cent). All other silicates may be considered present in trace amounts.

Amphibole Group

The amphiboles, particularly cummingtonite, actinolite, hornblende, and hornblende+cummingtonite, constitute the dominant mineral family in the Eastern Mesabi district. Minute amounts of riebeckite also occur. Richarz (1927) has previously studied some of the amphiboles occurring near Babbitt and the general Eastern Mesabi district.

Actinolite. The name actinolite, as applied in this report, refers to a variable group of amphiboles the optical properties of which lie between those of hornblende and cummingtonite. To the writers' knowledge, the properties of such a system are not available. The amphibole which Richarz (1927a, p. 700) studied earlier "differs from grunerite in the greater percentages of ferric iron and alumina, and from actinolite and common hornblende by the high amount of iron and the low percentages of lime and magnesia." This amphibole probably corresponds to what the writers call actinolite. Cummingtonite, hornblende, and actinolite were compared spectrographically and it was determined that although all contained moderate amounts of magnesium, hornblende and actinolite were appreciably richer in calcium and aluminum. Chemical analyses of the former two amphiboles show appreciable ferrous iron and moderate amounts of magnesium, but the calcium and the aluminum content of actinolite suggests that it may be more closely related to hornblende. It is considered better to use a more general name such as actinolite rather than attempt to propose a new name for the amphiboles. Tremolite, however, was not observed on the Eastern Mesabi range and therefore the use of actinolite implies the presence of members nearer the ferrotremolite end-member of the tremolite-ferrotremolite system rather than near the tremolite end-member, as the name usually implies. In comparison with hornblende and especially cummingtonite, actinolite is considered a subordinate silicate constituent of the iron formation.

In view of the preceding paragraph it is apparent that of a considerable variety or range of amphiboles, many are identified as actinolite. The dark green, pleochroic, low birefringent hornblende and the colorless, high birefringent cummingtonite are both easily recognized in thin section; in general all amphiboles observed in thin section that have various light shades of green and low to moderate birefringence are identified as actinolite. The generalized optical properties of the amphiboles called actinolite are listed below.

positive elongation (length slow) to amphibole cleavage
maximum extinction angle c$\wedge$Z about $14°$ ± 1
moderate to strong birefringence (ca. .020 to .025)
weak to moderate dispersion (r>v, rhombic and inclined)
polysynthetic twinning parallel to 100, uncommon

  optic plane 010   2V ca. $80\pm5°(-)$
  X<Y$\leqq$Z   X(ca. a) = pale green-yellow
  n_β' ca. 1.66   Y(= b) = medium green
  n_γ' ca. 1.68   Z(ca. c) = medium (bluish) green

In hand specimen dark green hornblende and pale yellow cummingtonite are readily recognized, but all pale green assemblages of amphiboles are collectively called actinolite in core logging. Thin sections of some of these specimens reveal that what was called actinolite in hand specimen actually is a combination of grains of hornblende and cummingtonite, or grains of actinolite and cummingtonite, or grains of hornblende+cummingtonite, as well as actinolite alone. These mineral assemblages are all closely associated in one place or another (see hornblende+cummingtonite). Since these assemblages cannot be distinguished with the hand lens, the reader must keep in mind that the distribution of actinolite described as occurring within the various submembers (see Chapter 2) can include any of the above assemblages.

Cummingtonite. With the exception of quartz, cummingtonite is the most abundant silicate in the metamorphosed iron formation, occurring in almost all submembers. As described in detail for each submember in Chapter 2, cummingtonite is not uniformly distributed throughout any given submember and cummingtonite varieties with different crystal habits can occur within a given member. The optical properties of these cummingtonite varieties are essentially identical, and are as follows:

positive elongation (length slow) to amphibole cleavage
maximum extinction angle c$\wedge$Z $= 16°$
strong birefringence (ca. .037)
weak dispersion (r>v, rhombic and inclined)
polysynthetic twinning parallel to 100 is common

  optic plane 010   2V ca. $85+°(-)$
  X$\leqq$Y = Z   X(ca. a) = pale yellow
  $n_\beta' = 1.658$ Y(= b) = pale yellow $\pm$ greenish tint
  $n_\gamma' = 1.677$ Z(ca. c) = pale yellow $\pm$ greenish tint

These optical properties indicate that this amphibole is cummingtonite, an intermediate member of the kupfferite-grunerite series (Winchell, 1951, p. 428). Results of a chemical analysis, discussed below, support this conclusion.

One variety of cummingtonite is commonly found in minor to moderate amounts only in the area south and east of Argo Lake. Here, the cummingtonite habit is medium- to coarse-grained, poikilitic and prismatic, and is generally associated with fayalite and poikilitic ferrohypersthene. This variety is referred to specifically as "poikilitic cummingtonite" in all descriptive text. In many places this type of cummingtonite appears to be essentially contemporaneous with the two last-mentioned silicates, but in other places it poikilitically encloses them (as well as hedenbergite) suggesting it formed slightly later, yet still within the same general sequence of crystallization.

The second, and most abundant, variety is the typically fine-to-medium-grained, prismatic to acicular cummingtonite which occurs in nearly all submembers of the iron formation. Unless otherwise specified (such as "poikilitic cummingtonite" or "tabular cummingtonite"), use of the mineral name cummingtonite with any other descriptive adjective refers to this most abundant variety. The majority of cummingtonite of this habit has developed in rock that previously consisted mainly of quartz and magnetite, as evidenced by abundant relic sedimentary structures still visible in the rock. This variety of cummingtonite is most abundant in the area immediately to the south and west of Iron Lake; it becomes somewhat finer-grained and acicular to the west, and is generally not abundant in the eastern holes.

Where it occurs with other silicates, the fine-grained, prismatic or acicular cummingtonite is found to have partly replaced fayalite, ferrohypersthene, poikilitic-prismatic cummingtonite, hedenbergite, and hornblende; there are no cases of the reverse relationship.

The third varietal habit of cummingtonite is uncommon and is distinguished by its dark gray-green color and tabular habit parallel to (100). In hand specimen it is similar to "hornblende," and except for a pale green color, has the optical properties of cummingtonite. X-ray powder photographs show that it has a unit cell essentially identical to that of cummingtonite. This variety of cummingtonite occurs only in the Upper Slaty member south of Argo Lake. The rocks in which it occurs generally are completely recrystallized, although layer-like zones of fine-grain magnetite within the coarse-grained mat of tabular crystals probably reflect initial bedding as seen in Figure 24.

A fresh sample of cummingtonite from exposures of submember L in the Mitchell pit was crushed, concentrated, and carefully handpicked under the binocular microscope prior to the chemical analysis given in Table 4.

The most significant feature of this analysis is the high MgO content, 13.66 per cent of the mineral. The presence of so much magnesium is difficult to explain by a process other than at least partial metasomatic addition because, according to analyses given by Gruner (1946, p. 10), none of the "primary" silicates contains more than 7.7 per cent MgO. It must also be remembered that, next to magnetite and quartz, cummingtonite is generally the most abundant mineral throughout the iron formation on the Eastern Mesabi, except of course where other metamorphic silicates have already developed. In many submembers of the Upper Slaty and Cherty members it locally exceeds 60 per cent of the rock. Some magnesium was present at the time of formation of the fayalite (MgO = 3.94%). Judging from the relatively uniform distribution of fayalite within some submembers, the writers presume this to represent an initial distribution of magnesium.

Petrographic data and field evidence indicate that mobile, chemically active, aqueous fluids were introduced into the iron formation during metamorphism and that they were able to serve as a medium and activating agent for the recombination of the existing taconite constituents into cummingtonite. The source of the water might have been from the "primary" hydrous silicates, or from connate water from the iron formation and adjacent sedimentary units, or a metasomatic addition from the many pegmatitic veins of the area — depending on the viewpoint held as to the origin of the metamorphic silicates.

The distribution of cummingtonite, previously described in detail for each submember in Chapter 2, is difficult to explain if one assumes that it was formed merely from recombination of the constituents of the "primary" silicates. In review, cummingtonite is typically most abundant in the area south and west of Iron Lake and much less abundant in the area nearer to the gabbro. Similarly, it is not commonly found near the larger pegmatitic veins but has normally formed just beyond the outermost occurrences of hedenbergite adjacent to the veins. In addition, fine-grained cummingtonite is most abundant in rock consisting of relatively fine-grained quartz and magnetite and is least abundant where fayalite, poikilitic ferrohypersthene, poikilitic cummingtonite, and hedenbergite occur. The writers cannot visualize such a unique initial distribution of magnesium that is so fortuitously zoned around the gabbro and pegmatite contacts, nor can they explain why any available magnesium would not have been incorporated into anhydrous silicates during the earlier stages of metamorphism as has apparently happened elsewhere during the formation of fayalite and ferrohypersthene. If, however, metasomatic activity (mainly the addition of water and magnesium) was initiated shortly after the beginning of the thermal metamorphism produced by the gabbro, then the unique distribution and the magnesium-rich composition of cummingtonite is simply and easily explained (see Chapter 4).

Many specimens from most submembers of the iron formation show an intimate association of cummingtonite with hornblende, hornblende+ cummingtonite, and actinolite, as described elsewhere in this chapter.

This association, and the textures displayed by these minerals, suggest the existence of a discontinuous series or group of amphiboles that is intermediate between the kupfferite-grunerite and tremolite-ferrotremolite systems. In a given specimen it appears that hornblende could form as long as sufficient amounts of calcium and aluminum were available and that when these constituents were consumed, or were not continuously supplied, then cummingtonite continued to grow over the hornblende lattice or proceeded to form alone. Minor amounts of cummingtonite and actinolite have formed, in a retrograde fashion, from the hornblende present in some pegmatite veins. Cummingtonite is also found intimately associated with oligoclase, green and brown biotite, apatite, calcite, and zircon. Minute zircon grains are commonly enclosed in these cummingtonite grains and have produced small pleochroic haloes in the cummingtonite.

In summary, its unique distribution and magnesium-rich composition; its intimate association with the metasomatic calcium-rich silicates, biotite, apatite, zircon, and feldspar; its partial to complete replacement of all other metamorphic ferrous silicates; its greatest abundance in rock consisting essentially of fine-grained quartz and magnetite, all strongly suggest that cummingtonite — as it occurs on the Eastern Mesabi range — is, in part, metasomatic.

Hornblende. An important mineral constituent of the pegmatitic veins that occur on the Eastern Mesabi range, hornblende is also locally abundant in the iron formation adjacent to the veins. Its content within the pegmatitic veins, or even within a given vein, ranges from zero to nearly 100 per cent, and in these veins it is typically associated with alkali feldspar (albite), green biotite (and secondary chlorite), quartz, calcite, apatite, and sulfides, especially pyrite and some chalcopyrite. Hornblende within the veins and in the adjacent iron formation has the same optical properties and identical cell dimensions as shown by x-ray powder photographs. It is characteristically coarse- to extremely coarse-grained (one broken crystal found was 6 inches wide) within the pegmatites and is typically medium- to coarse-grained where it has formed in the iron formation adjacent to the veins.

The optical properties of the hornblende are as follows:

positive elongation (length slow) to amphibole cleavage
maximum extinction angle $c \wedge Z = 12°$, local dispersion effects
moderate birefringence (ca. .018), slightly masked by color
moderate to strong dispersion ($r > v$, rhombic and inclined)
simple twinning parallel to 100 in large crystals

optic plane 010 $2V$ ca. $75°(-)$
$X < Y \leq Z$ X(ca. a) = medium green-yellow
$n_\beta' = 1.678$ $Y(= b)$ = olive green
$n_\gamma' = 1.694$ Z(ca. c) = bluish green

A single crystal of hornblende from a pegmatite vein was crushed, freed of inclusions, and then submitted for chemical analysis (Table 4).

Pegmatites occur in the Upper Cherty member in the Mitchell pit exposures, and cores of other members from holes throughout the area also contain minor amounts of hornblende. Thin sections of these specimens show that it is generally intimately associated with feldspars, calcite, hedenbergite, actinolite, and hornblende+cummingtonite. Wherever found it is assumed — on the basis of occurrence and associations elsewhere — that its origin is, in part, metasomatic. The common occurrence of pleochroic haloes around small zircon grains appears to support this assumption.

Hornblende commonly appears with quartz in the interstitial matrix of the granulose fayalite-bearing hornfelses of the Lower Slaty member. It also has replaced, or has poikilitically enclosed, hedenbergite and the relatively coarse-grained poikilitic ferrohypersthene and cummingtonite that locally appear with fayalite. In other places it has clearly replaced hedenbergite and has, in turn, been replaced by, or is contemporaneous with, finer-grained actinolite, hornblende+cummingtonite or cummingtonite. In many specimens it appears to be slightly altered to nontronite in a manner similar to that noted for actinolite and hedenbergite. In numerous specimens stilpnomelane is interstitial to and has partly replaced hornblende both within the pegmatite veins and the adjacent iron formation. In some of the pegmatite veins, hornblende has been altered and bleached to a creamy white, fine-grained material that still retains an amphibole-like structure.

Hornblende+Cummingtonite. Hornblende+cummingtonite is a collective mineral name used by the writers to indicate the presence of two distinct amphibole phases within a given grain or assemblages of grains; the phases consist of cummingtonite and of hornblende and (or) actinolite. This "composite amphibole" is not a mechanical mixture of two types of amphibole grains but consists of two separate phases in homoaxial intergrowth (the optical orientations of the phases being nearly parallel), in which part of a given grain has the optical properties of cummingtonite (colorless, highly birefringent, polysynthetically twinned) and the remaining part of the same grain has the properties of, for example, hornblende (green and strongly pleochroic, low birefringence, and untwinned). Examination of these grains in plane polarized light shows no physical discontinuity between these areas except color, and it is clear that within a given grain the prismatic cleavage and the optic plane (010) are geometrically the same for each phase. Actually any two of the three minerals — hornblende, actinolite, or cummingtonite — can show a similar relationship. Those grains consisting in part of hornblende and cummingtonite are most conspicuous, but the minor occurrences of those grains consisting of actinolite and cummingtonite might be slightly more numerous. Although there are many exceptions, the cummingtonite portion of a given grain generally constitutes the larger and outermost parts of the grain. Richarz (1927b, p. 153) has noted some specimens from Michigan in which small pleochroic blue-green patches of lower birefringence occur within the usually colorless, highly birefringent grunerite. His descriptions seem to refer to material that would be called hornblende+cummingtonite by the writers.

Grains of hornblende+cummingtonite constitute perhaps only a few tenths of a per cent of the amphibole grains present in the Eastern Mesabi, yet this composite mineral typically occurs in minute to trace amounts in nearly every submember of the iron formation. It most frequently occurs in many cummingtonite-rich specimens. The most interesting and probably significant occurrence is within, or adjacent to, thin magnetite lamellae along, or between which, calcium-bearing solutions appear to have moved during metamorphism. In specimens of this type one can observe hornblende+cummingtonite occurring essentially within a transition zone between a hornblende- and (or) actinolite-rich area and a cummingtonite-rich area, as shown in Figure 30.

These transition zones, consisting of the mixed-grains of hornblende+cummingtonite can occur laterally within a given group of lamellae or else within the granules, adjacent to the lamellae, which consist of, for example, hornblende near to the lamellae and cummingtonite farther away and, incidentally, mostly quartz and magnetite beyond this zone of local metasomatism. In many specimens any or all of the amphiboles, especially hornblende, contain strong pleochroic haloes that have developed around zircon; feldspars are also commonly present in these specimens.

The textures or fabrics of assemblages containing hornblende+cummingtonite within cummingtonite-rich specimens suggest that when the range of stable equilibrium for the formation of amphiboles was reached, hornblende and (or) actinolite began to form locally along with cumming-

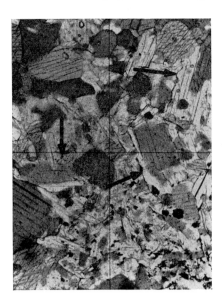

FIGURE 30. — Hornblende+cummingtonite. Commonly occurs along the transitional zone between hornblende-rich parts of taconite (left) and the cummingtonite-rich rock (right). The compound mineral name alludes to grains consisting of homoaxial intergrowths of hornblende, commonly in the central part of the grain, and cummingtonite, the outer part of the same grain (arrows). A few typical pleochroic haloes can be seen in the large hornblende grain (lower left). (32-199½; 100X)

tonite and continued to form while sufficient amounts of calcium and aluminum were available, and when these constituents were exhausted cummingtonite continued to develop upon the earlier amphibole structure with no obvious discontinuity. Similarly, the textures or fabrics of the amphiboles within or adjacent to the magnetite lamellae suggest that hornblende and (or) actinolite began to form locally and continued to form as long as the chemically active fluids which moved along the lamellae continued to supply the constituents necessary for their formation. Beyond this point, exhaustion or consumption of these constituents permitted only cummingtonite to form. Hornblende+cummingtonite, of course, is found in the transition zone between the hornblende- and cummingtonite-areas and it probably formed in a manner similar to that described at the beginning of the paragraph.

Riebeckite (Soda-Hornblende). Prismatic riebeckite was found only in submember I of hole 35 and submember J of hole 28 where it clearly replaced relic magnetite-quartz granules, as shown in Figure 19. It may be easily distinguished from other amphiboles of the area by its optical properties which are as follows:

negative elongation (length fast) to amphibole cleavage
maximum extinction angle of 8° (incomplete extinction)
abnormal (tan to violet) low-moderate birefringence (ca. .015)
extreme dispersion (r << v, horizontal)

optic plane + 010 $2V$ ca. $90°(-?)$
$X >> Y < Z$ Y(ca. a) = light (greenish) tan
$n_\beta' = 1.671$ $Z(= b) = $ pale blue to violet
$n_\gamma' = 1.674$ X(ca. c) = light (greenish) blue

These properties indicate that it is an intermediate member of the glaucophane-riebeckite series (i.e., crossite) lying close to the riebeckite end-member in composition (Winchell, 1951, p. 441). It is herein referred to as riebeckite for convenience in lieu of the less commonly used name crossite.

White (1954, p. 64) identified the fibrous variety, crocidolite, in taconite near the Aurora sill, and concluded from his field studies that soda from the sill combined with the constituents of the taconite to yield the mineral. The writers have no field or petrographic data to indicate the source of the soda for the minor amounts of riebeckite on the Eastern Mesabi; it is probably metasomatic, at least in part. Its relative age with respect to the other silicates on the Eastern Mesabi was not determined.

FELDSPAR GROUP

Although it is recognized that albite is also a plagioclase feldspar, this mineral is referred to in the text as alkali feldspar, which helps to distinguish it from subordinate amounts of other plagioclase feldspar occurring with it or separately.

Alkali Feldspar (Albite). Alkali feldspar is a major constituent in most of the larger pegmatitic veins and is locally abundant in the adjacent wallrock. An x-ray diffraction powder photograph and infrared spectrophotometry reveal a structure very similar to albite but distinctly different from either orthoclase or microcline. A spectrographic study shows that the mineral is deficient in potassium but contains moderate amounts of calcium, some barium, and minor strontium. Although the spectrographic plate is not sensitive to the sodium "D" lines, a simple spectroscopic test shows the feldspar to contain sodium. On the basis of this data, the alkali feldspar is probably an impure calcium-barium-strontium-bearing albite.

In general, the amount of alkali feldspar in the wallrock increases directly with the amount in the pegmatites. It is most abundant in the pegmatites near the gabbro. Unlike plagioclase found in the wallrock, it is commonly very coarse- to coarse-grained and is sufficiently abundant to be evident in the iron formation next to the veins. In a few exceptional cases the taconite of the Lower Cherty member has been greatly metasomatized and reconstituted into a coarse-grained, oligoclase-magnetite-quartz-hornblende-alkali feldspar crystalline rock. It commonly occurs, however, in small (up to 4 inches) sill-like masses, that are also usually rich in hornblende, or in small pegmatoid-like pods or segregations composed of minerals found in the nearby pegmatitic veins. The sill-like bodies and segregations are abundant locally near the pegmatites and diminish in size and number away from them, though some alkali feldspar has been found as far as 150 feet distant.

In thin section the alkali feldspar is cloudy, of pinkish tint, poorly twinned (albite-pericline?), and has a negative relief with all indices between 1.53 and 1.54 and a $2V = 85°(+)$. It is found to be associated with any or all of the metasomatic minerals, and actinolite, hornblende+cummingtonite, and chlorite have replaced the feldspar at several places. In most places it is slightly altered to sericite.

Plagioclase (Oligoclase and Andesine). Plagioclase is generally subordinate to alkali feldspar (albite) in the pegmatitic veins but is normally more abundant within the iron formation than is alkali feldspar. An exception to this generalization can be made for the rock adjacent to some of the larger alkali feldspar-rich pegmatites.

The optical properties of the plagioclase within the iron formation show that it is oligoclase of composition, ca. An_{15}. Measurement of the indices of refraction of the plagioclase of the pegmatitic veins indicates andesine, ca. An_{35}. Plagioclase grains within the metamorphosed iron formation are generally fine-grained, anhedral, clear and colorless, and poorly twinned according to the albite law. It is most typically associated with the calcium-bearing amphiboles, cummingtonite and brown biotite. Within such assemblages the iron silicates commonly contain pleochroic haloes around minute grains of zircon.

It is assumed that the formation of oligoclase within the iron formation is related to the metasomatic activity accompanying the emplacement of the pegmatitic veins.

Garnet Group

Both iron-aluminum garnet (almandite) and calcium-iron garnet (andradite) are present in the iron formation, but only the latter is locally abundant.

Almandite. This garnet is rare in the district. Several small subhedral grains are present in submember S from holes 32 and 21 where they are associated with fine-grained biotite. In submember R from hole 21, many large subhedral grains of almandite have been incipiently replaced by cummingtonite and biotite, as shown in Figure 10. This garnet also occurs as small dark-gray to black anhedra that are difficult to observe, especially in the "Intermediate Slate" horizon.

Andradite. Brownish-red andradite, the most common garnet, is a characteristic mineral in core from submember G and to a lesser extent in units F and I, especially in holes west of Iron Lake. It is rarely found in core of other members. Pit exposures show locally abundant amounts throughout the Upper Cherty member, but it is essentially absent in core from the same area. In the Mitchell pit, andradite is most abundant near the gabbro and adjacent to pegmatitic veins. In these environments, the garnet developed along a given horizon and possibly reflects the initial compositions of the beds. An alternate possibility is that the garnet is in part metasomatic and that beds of relatively greater permeability have determined their loci.

Andradite develops within small ($\frac{1}{2}''$ to $12''$) pegmatoid-like pods in quartzose taconite and is commonly associated with calcite, epidote, and quartz, and to a lesser extent with hedenbergite, magnetite, and diopside. Optical properties and x-rays show the garnet to be andradite, and spectrographic study demonstrates that calcium, iron, and aluminum are essential constituents.

Mica Group

Muscovite is rare in all rock units on the Eastern Mesabi district, but both green and brown biotite are common in minor amounts in many of the rock units.

Biotite. In addition to being a prominent mineral constituent of the older and younger meta-sedimentary formations and igneous rock units of the Eastern Mesabi district, biotite is locally common in the Biwabik iron-formation. Both green and brown varieties are present, but neither is restricted to specific stratigraphic horizons of the iron formation because their distribution seems to depend primarily upon the presence or absence of the metasomatic activity that was probably necessary for their formation.

Large tabular crystals of green biotite, up to 1 inch diameter, are found to be locally abundant in the pegmatitic veins exposed in the Mitchell pit. Microscopic and x-ray diffraction study shows that much of this biotite has been altered to green chlorite. Several thin, green, biotite-rich, cummingtonite-bearing, layer-like masses within the metasomatized adjacent wallrock have been traced outward from the pegmatite veins until they thin out and disappear. It seems apparent that the biotite and cummingtonite of these layers is, at least in part, metasomatic. It also commonly occurs with the feldspars and metasomatic calcium-rich silicates in the regions adjacent to the pegmatites. This association, plus the fact that it commonly contains minute grains of radioactive zircon(?) that have produced strong pleochroic haloes within it, appears to support a metasomatic origin. Among the silicates it is mostly associated with the relatively fine-grained actinolite and cummingtonite, as repeatedly mentioned in the descriptive stratigraphy of the formation. Where these occur together, both commonly contain pleochroic haloes about minute zircon crystals.

Because they commonly occur together in differing amounts, there appears to be no systematic pattern for the distribution of green or brown biotite. It is apparent, however, that the green biotite is often associated with the calcium-rich silicates while the more common brown biotite is closely associated with the magnesium-rich cummingtonite.

Muscovite. Although minute amounts of muscovite have been found in some of the siliceous pegmatite veins it is not a typical pegmatite mineral in this district, the most common mica in the pegmatites and adjacent taconite being biotite. In a specimen of submember K from hole 35, minor amounts of muscovite have formed essentially interstitially to the quartz grains both within the granule-like structures and the surrounding matrix, and are also closely associated with minor amounts of riebeckite. The micaceous grains are small and scattered, and identification has been based upon the variation of refractive index with orientation and upon their birefringence.

Pyroxene Group

Pyroxenes are an important group of silicate minerals of the Eastern Mesabi, second in abundance only to fayalite and the amphiboles.

Clinohypersthene. A pyroxene, tentatively identified as clinohypersthene, is found only in specimens of submember I from holes 28 and 30. The mineral is pale straw-yellow in color, bladed to prismatic in habit, and resembles hypersthene in hand specimen but under the microscope it is colorless, has an extinction angle of $c \wedge Z = 38°$, $2V = 55°$ (+), weak dispersion and a pyroxene cleavage. Although the birefringence of the mineral is low ($B = 0.015$), other optical properties seem to confirm its identification (Winchell, 1951, p. 409). The high and low indices of cleavage fragments of the mineral are greater and smaller than oil of index $n = 1.738$. It is possible that this pyroxene may be an intermediate member of the discontinuous series between the clinoenstatite and diopside-hedenbergite systems even though the extinction angle appears to be about 10° too small. The x-ray powder photograph is similar to some available x-ray film standards of pyroxenes but did not exactly match any of them.

The mineral is intimately associated with subordinate amounts of hornblende, actinolite, calcite, plagioclase, and minute amounts of hedenbergite. Its close association with these minerals indicates that it is of metasomatic origin.

Diopside. Diopside occurs to a minor extent in submember A and especially in B where it is locally abundant along with moderate amounts of hedenbergite. In a few places it occurs in minor amounts with andradite present in other submembers. In contrast to dark green hedenbergite, diopside has a characteristically light grayish-green color in hand specimen, but it is less easy to distinguish between them in thin section. The most obvious differences are that hedenbergite is greener in thin section and has an extinction angle (c$\wedge$Z) of 48 degrees while diopside is very pale green and has an extinction angle of 42 degrees; the optic angles, optic signs, and dispersions are nearly identical. The minerals can be readily recognized and distinguished by examining crushed grains in an oil of refractive index 1.720; all indices of diopside are lower and all indices of hedenbergite are higher. X-ray powder photographs also serve to distinguish between these pyroxenes.

Diopside might have formed during thermal metamorphism if magnesium-bearing constituents were initially present in unit B. On the other hand, magnesium might have been introduced from nearby pegmatites.

Hedenbergite. The most abundant pyroxene in the iron formation of the Eastern Mesabi range, hedenbergite is a common mineral constituent in many core specimens of all submembers except units A and Q. The distribution of hedenbergite within the core does not form an easily recognizable pattern. Numerous occurrences of hedenbergite in the Mitchell pit, however, show a very obvious and simple pattern of distribution in the wallrock adjacent to pegmatitic veins, as discussed in Chapter 2. Its occurrence, abundance, and grain size are clearly not related to proximity to the Duluth gabbro except that the pegmatites are somewhat more numerous near the gabbro. Many specimens from holes along the western margin of the area contain as large, or greater, amounts of coarse-grained hedenbergite as those from holes near the gabbro.

The main optical properties of hedenbergite are as follows:

maximum extinction angle c$\wedge$Z = 48°
moderate birefringence (ca. .025)
weak to moderate dispersion (r>v, rhombic and inclined)

optic plane 010	2V ca. 60°(+)
X = Y = Z	light green
$n_\beta' = 1.724$	non-pleochroic
$n_\gamma' = 1.736$	Y = b

A specimen of hedenbergite, obtained from the wallrock adjacent to a pegmatite, was crushed, concentrated, and analyzed chemically (see Table 4).

From the high content of calcium in the mineral, as well as from the well-defined distribution and great abundance of hedenbergite adjacent to the pegmatitic veins, it is obvious that the veins were sources of metasomatic activity from which appreciable amounts of calcium, and some magnesium, were added to the adjacent rock.

Hedenbergite is most commonly developed within taconite consisting mainly of quartz and magnetite and, judging from partly replaced relic sedimentary textures, has formed directly from these minerals upon addition of sufficient calcium and magnesium during metamorphism. Where fayalite- or ferrohypersthene-bearing assemblages are also present, the hedenbergite has always poikilitically enclosed or partly replaced these minerals. Locally, hornblende, actinolite, and hornblende+cummingtonite appear to be contemporaneous with hedenbergite; as a rule, however, they have poikilitically enclosed or have partly replaced it. A few specimens of hedenbergite show fabrics that strongly suggest incipient retrograde alteration to fine-grained actinolite. Fine-grained prismatic to acicular cummingtonite does not commonly occur with hedenbergite in the relatively coarse-grained quartz-magnetite-rich taconites adjoining the pegmatites. But wherever they occur together, hedenbergite has always been replaced by, and commonly appears only as small remnants within, the cummingtonite.

In many of the specimens studied, hedenbergite is moderately altered in the neighborhood of hisingerite veinlets. In several places small masses of nontronite-rich material, pseudomorphous after hedenbergite, have formed close to hisingerite veinlets that have also been altered to nontronite. This seems to be the result of low-temperature hydrothermal (auto?) alteration.

Ferrohypersthene. Although rarely an abundant mineral in the Lower Slaty and the bottom of the Upper Cherty members in most of the holes east of Iron Lake, ferrohypersthene is nevertheless locally common. The distribution of this fine- to medium-grained pyroxene appears to be somewhat parallel to the present gabbro contact.

In the Mitchell pit, medium- to coarse-grained ferrohypersthene is also found in many places as rounded clusters of anhedra in the wallrock adjacent to the pegmatitic veins. The grain-size and abundance of the pyroxene decreases with distance from the pegmatite, having formed either by recombination of original taconite material or more probably during partial metasomatism by magnesium-bearing fluids from the pegmatitic veins.

A large cluster of coarse-grained anhedra (up to 2 inches in diameter) collected from the Mitchell pit was crushed, purified, and then chemically analyzed (Table 4).

The optical properties of ferrohypersthene are as follows:

positive elongation (length slow) to pyroxene cleavage
straight and symmetrical extinction
moderate birefringence (ca. .017)
weak to moderate dispersion (r<v, rhombic)

optic plane 010 2V ca. 65°(−)
X ≥ Y ≤ Z   X(= a) = pale pinkish yellow
$n_\beta' = 1.728$ Y(= b) = pale yellow (pinkish tint)
$n_\gamma' = 1.750$ Z(= c) = pale greenish yellow

On the basis of these optical properties, and the results of the chemical analysis, it is clear that the mineral is ferrohypersthene according to the classification proposed by Winchell (1951, p. 406).

OTHER SILICATES

Chlorite. Green chlorite, secondary after green biotite, is most common in the pegmatitic veins and is locally abundant in the Pokegama formation and the Giants Range granite. Within the iron formation, it occurs in assemblages containing especially feldspar, apatite, biotite, and actinolite: minerals which have been interpreted to be partly metasomatic. In many places it appears to be secondary after biotite because of its retention of pleochroic haloes that locally characterize biotite. In other places it is essentially interstitial to the abundant minerals of the iron formation and there is no petrographic evidence to indicate whether it is secondary or not. On the Eastern Mesabi, chlorite is generally present in minor to trace amounts in the Biwabik iron-formation, and in all occurrences it appears to have formed during the waning or more hydrous stage of the general metamorphism.

Epidote. Although not common, epidote is often associated with many of the irregular pegmatite-like pods of andradite-quartz-calcite in the Mitchell pit exposures of the Upper Cherty. Core specimens of submembers E of hole 25, F of hole 35, and G of holes 19 and 32 contain minor amounts of epidote associated with andradite.

Fayalite. A common mineral constituent within the lower part of the Upper Cherty and the upper part of the Lower Slaty members, fayalite is also present to a minor extent in other members of the formation. Beginning with submember M of the Upper Cherty member, the fayalite content generally increases to a maximum in submembers O and P, and decreases to minor amounts near the bottom of the Lower Slaty member. In general, the average grain size and abundance of fayalite is greater near the gabbro and diminishes to the west away from the gabbro. To the south and west of Argo and Iron lakes, fayalite occurs as small remnants within fine-grained cummingtonite; a few of these remnants are present in submember Q as far west as hole 33.

A fayalite-rich specimen from submember M in the Mitchell pit was crushed and the fayalite concentrated and chemically analyzed (Table 4). The optical properties are as follows:

positive elongation (length slow) to 010 cleavage
straight extinction
strong birefringence (ca. .049)
moderate dispersion (r>v, rhombic)

optic plane 001
2V ca. 55°(—)
X ≤ Y ≅ Z
all indices > n = 1.785

Z(= a) = pale yellow
X(= b) = pale yellow
Y(= c) = pale yellow to brownish tint

On the basis of the chemical analysis and the observed optical properties, the mineral lies between ferrohortonolite and fayalite in composition. Because it is near the fayalite end-member, the name fayalite is used.

Fayalite appears in the earliest assemblages formed during metamorphism. Poikilitic ferrohypersthene and cummingtonite locally enclose fayalite and thus appear to have formed slightly later in the same general sequence of recrystallization. Hedenbergite and hornblende have also locally replaced or poikilitically enclosed fayalite. In some places where fayalite occurs as remnants within cummingtonite, trace amounts of magnetite and (or) an unidentified green serpentine-like mineral (possibly hisingerite) locally occur along the contact between the minerals. Wherever cut by hisingerite veinlets, fayalite is commonly altered in part to brown or green hisingerite adjacent to the veinlets.

Hisingerite. One of the last silicates to form in the iron formation on the Eastern Mesabi range, hisingerite occurs in thin (up to ¼ inch max.) veinlets that are found in all members of the iron formation. Hisingerite, however, is most common in the area south of Iron Lake and is locally abundant in submembers M, N, O, P, and Q.

Most of the thin veinlets consist of a greenish-black hisingerite, bright emerald green in thin section, with some magnetite. In some veins, up to 1 inch thick, coarse-grained biotite and quartz and very coarse-grained albite occur with the hisingerite. Some veins also contain minor amounts of dark brown hisingerite, orange brown in thin section, which might be a more ferric variety. Black hisingerite was removed with a small chisel and hammer (the minor amounts of the brown variety were rejected) from a large sample of a "fresh" greenish-black veinlet collected from submember M near the eastern end of the Mitchell pit. The chemical analysis of hisingerite is given in Table 4 and its optical properties are as follows:

amorphous and, rarely, incipiently crystallized
n = 1.63 to 1.64 and emerald green where fresh (ferrous?)
n = 1.65 to 1.69 and orange to brown where altered (ferric?)
(incipiently crystallized varieties are very weakly birefringent and pleochroic)

In thin section, hisingerite veinlets are found cutting through fayalite, poikilitic ferrohypersthene and cummingtonite, hedenbergite, hornblende, fine-grained cummingtonite, and quartz. As a general rule fayalite is greatly altered to green and (or) brown hisingerite where it is adjacent to the veins, and the other minerals are successively less altered in the approximate order given above; cummingtonite and quartz are generally unaltered. Within a given assemblage cut by a hisingerite veinlet, it is generally true that the most anhydrous mineral present is the one that is most altered by the hisingerite-forming fluids. Some silicates, notably hedenbergite, hornblende, and actinolite (i.e., the calcium-bearing ferrous silicates) are commonly altered to nontronite.

Idocrase (Vesuvianite). Idocrase occurs only in submember A from hole 21 as small anhedra essentially interstitial to calcite grains and is interpreted to be a product of contact metamorphism by the Duluth gabbro upon impure primary limestone. The optical properties of high relief and low birefringence suggest idocrase and this was verified by x-ray analysis.

Minnesotaite. This mineral was observed only in drill core from hole 27E where it is an important constituent of granules, chert layers (now quartz), and magnetite-rich layers and lamellae. It is commonly associated with calcite, ankerite and (or) siderite.

Petrographic data indicate that minnesotaite is relatively later than the relic sedimentary textures preserved in the rock, but there is nothing to indicate whether it formed under low temperature hydrothermal conditions or is diagenetic. The writers favor the concept that minnesotaite is a metamorphic mineral, and is probably the low-temprtature, more hydrous equivalent of cummingtonite; however, until these two minerals can be found together in a specimen this cannot be verified. The most interesting fabric involving minnesotaite is shown in Figure 31. Here, the

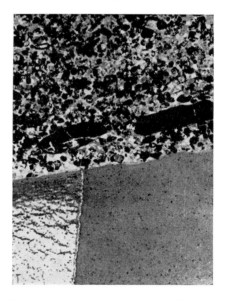

FIGURE 31. — Minnesotaite. Initially the area shown consisted of a chert layer and a quartzose layer containing numerous magnetite-rich granules. The chert layer was subsequently cut by a carbonate-quartz veinlet and numerous very small rhombohedra of carbonate (small dark spots and streaks) formed in this layer. Subsequent to this, mobile minnesotaite-forming solutions permeated the chert layer from the right. The central veinlet essentially acted as an impermeable barrier and the minnesotaite (light gray) was precipitated mainly to the right of this solution barrier. Minnesotaite replaces carbonate grains, magnetite, and quartz in this specimen. The magnetite-quartz granules of the layer above are moderately recrystallized and contain only minor amounts of minnesotaite. (27E-195½; 3X)

thin chert and adjacent granule-bearing layers were deposited and were sufficiently indurated to sustain small fractures. Carbonate-lined quartz veinlets then filled these fractures, although an alternate hypothesis might be that they were replacement veinlets. It appears as if minnesotaite-forming solutions could not effectively penetrate the impermeable carbonate-quartz veinlets and were thus able to precipitate minnesotaite only to the right of the barrier. The mineral fabrics indicate that mobile minnesotaite-forming fluids were once active in this specimen. These fabrics, of course, do not indicate whether these fluids were active during diagenesis or metamorphism at a much later date. Although the absence of proof to the contrary does not entirely justify a conclusion, the writers favor the latter alternative.

In thin section the very fine-grained acicular cummingtonite present in core from holes near Rangeline 14W is similar to minnesotaite, but the relief of the grains does not vary greatly with rotation of the microscope stage. X-ray diffractograms of these specimens do not show the presence of minnesotaite.

Nontronite constitutes a minute fraction of the silicates present on the Eastern Mesabi range. Its occurrence is not restricted to any particular member of the iron formation but is typically secondary after hisingerite in veinlets. Many of these veinlets have been partly or completely altered to nontronite. Although some field relations suggest surface weathering, it might be the product of late-stage auto-alteration brought about by the spent solutions which deposited the hisingerite. The observation that fine- to coarse-grained pyrite cubes are commonly found in the altered veinlets now consisting largely of nontronite but are never found in the fresh hisingerite veinlets probably lends support to the latter view.

In many places where hisingerite veinlets exist, the adjacent calcium-bearing ferrous silicates — hedenbergite, actinolite, and hornblende — have been altered to nontronite. In these examples it appears that the hisingerite-forming fluids permeated the adjacent wallrock, altering the silicates to nontronite. In most thin sections this alteration is incipient and has given a brownish color to the silicates. In some specimens alteration is nearly complete and the product, nontronite, has optical properties similar to those of stilpnomelane.

Quartz. The most abundant mineral of the iron formation is quartz. Relics of "chert" granules and distinctly layered and laminated "chert" beds are the most apparent structures preserved in the now quartzose material of the iron formation of the Eastern Mesabi range. Relic structures of this type from almost all of the submembers suggest that it was initially precipitated as chert, although clastic grains (originally either chert or quartz) are visible in many places. Most of the relic structures and textures are defined or outlined by disseminated fine-grained magnetite in fine-grained quartz.

Quartz also occurs as anhedra within the granoblastic fayalite- and poikilitic ferrohypersthene and cummingtonite-bearing assemblages. In some places these assemblages are partly replaced by prismatic and acicular cummingtonite-quartz assemblages.

Quartz-magnetite assemblages commonly have been replaced by hedenbergite, hornblende, hornblende+cummingtonite, actinolite, cummingtonite, and to a lesser extent by calcite, hisingerite, as well as stilpnomelane.

Near the gabbro, some recrystallized quartz grains attain diameters up to 6 or 8 mm, although diameters of ½ to 1 mm are closer to the average for extensively recrystallized quartz in the cores from the eastern holes. Except in a very general way, the grain size of quartz is not a reliable indicator of proximity to the gabbro because of the influence of the many pegmatitic veins that cause local recrystallization of the nearby quartz. Almost all of the quartz is fine-grained in the western holes and is rather uniform in size, normally ranging between .08 and .4 mm in diameter.

Sphene (Titanite). Within the iron formation sphene was found in only two holes. In submember B of hole 21, it is intimately associated with an assemblage of diopside, apatite, quartz, actinolite, and plagioclase. Minor amounts adjacent to a diabase sill were also found in hole 32. As previously described, the sill has been metamorphosed and along the contact with the iron formation some sphene has formed, the titanium presumably having come from augite within the sill.

Stilpnomelane. In contrast to the Main Mesabi range, stilpnomelane is present only in trace amounts on the Eastern Mesabi range. In thin section it has optical properties similar to those of nontronite and biotite except for birefringence. Its birefringence is greater than that of nontronite and somewhat smaller than for biotite.

Stilpnomelane occurs to a minor extent in almost all of the calcium-rich pegmatitic veins, where it has formed interstitially to, or has partly replaced, hornblende; it has also partly replaced hornblende in the wall-rock adjacent to some large pegmatites. Minor amounts are commonly associated with calcite, quartz, and rarely barite and plagioclase, in thin veinlets that cut the iron formation. Some specimens containing these veinlets also contain small laths of stilpnomelane in fine-grained magnetite-quartz granules and fine-grained quartz layers where it has partly replaced the sedimentary structures. In a few of these specimens stilpnomelane has formed in a retrograde manner after actinolite and hornblende, which in some places is secondary after diopside and hedenbergite. In the light of these textures, the limited range of thermal stability of stilpnomelane, and the temperatures of formation of the ferrous silicates, stilpnomelane on the Eastern Mesabi range has clearly formed as a retrograde mineral following the development of the abundant metamorphic silicates rather than persisting as relic diagenetic stilpnomelane.

In hole 27E, stilpnomelane is a common mineral within granules, chert beds and magnetite-rich lamellae, and thin layers. In these occurrences, however, there are no petrographic data to indicate whether the stilpnomelane formed under conditions of low temperature metamorphism (which may or may not have been related to the emplacement of the gabbro) or during diagenesis shortly after deposition, because grain fabrics could be identical under either condition. There is a lack of any conclusive petrographic texture that would indicate either of these possibilities; the only thing that is apparent is that the stilpnomelane is later than the relic sedimentary structures. Adjacent to some magnetite lamellae, stilpnomelane-rich granules have formed both within and adjacent to the lamellae. Slightly farther away the granules are only partly composed of stilpnomelane, and fine-grained quartzose granules with minute amounts of stilpnomelane lie farthest from the lamellae, as shown in Figure 11. This texture could have been produced by extensive diagenetic activity. Similar textures are also displayed by hornblende, actinolite, and other minerals on the Eastern Mesabi, clearly the result of metasomatic activity. In both cases it seems obvious that the solutions moving within the lamellae have permeated the surrounding area to form the appropriate minerals, but it is impossible to determine from petrographic data the date of these events.

Stilpnomelane is commonly associated with quartz and calcite veinlets, or less commonly with ankerite and siderite, that cut the formation in core from hole 27E. In hole 27E, the veins are generally stilpnomelane-rich, locally pyrite-bearing, and follow the bedding in many places. Stilpnomelane has locally formed in the wallrock adjacent to the veinlets and is believed to have been formed by fluids emanating from the veinlets. There is no evidence indicating whether the carbonate-stilpnomelane-quartz veinlets on the Eastern Mesabi range are the same relative age as those encountered in hole 27E near Mesaba. They are obviously later than the formation of the higher grade metamorphic silicates in the eastern district, and unless those present in the region near Mesaba were formed at essentially the same time, although probably under conditions of lower temperature, then recurring solutions of two different generations are necessary to explain their occurrences. This, of course, is possible, but the question remains, "Is it probable?" In the absence of evidence to the contrary, it is assumed that the simplest explanation holds and they are of the same relative age. This indicates that stilpnomelane is a low temperature metamorphic mineral in the iron formation, which is a fact on the Eastern Mesabi range but is only implied for the region near Mesaba to the west.

Tourmaline. Tourmaline is rare in the iron formation and was found only near the contact with the Pokegama formation in hole 32. Both in the Biwabik and Pokegama formations, it is intimately associated with cummingtonite, green and brown biotite, quartz, and actinolite, and appears to have formed contemporaneously with them. It possesses the typical tourmaline absorption formula $O > E$ and has a pleochroic formula of E = pale violet, O = dark slate blue. On the basis of its color and moderate birefringence (ca. 0.025), it is identified as the iron tourmaline, schorlite.

Wollastonite. Found only in submember A of hole 32, wollastonite occurs there as elongate (up to 1 inch) prismatic grains within a fine-grained lime-silicate hornfels. From metasomatic mineral assemblages in the adjacent submember, it seems clear that the thermal effects accompanying the emplacement of a pegmatite vein were responsible for the formation of the wollastonite.

Zircon. Zircon has been tentatively identified as minute radioactive grains that produce the small pleochroic haloes in the amphiboles and biotites found in the iron formation and the pegmatitic veins. Identification is made on the basis of shape, very high relief, strong birefringence and optical orientation (length slow, straight extinction). The assumption is made that all the minute blebs of radioactive minerals causing pleochroic haloes in the amphiboles and biotites are zircon, although it is possible that other accessory minerals such as xenotime, monazite, or allanite are also present in this form.

4. METAMORPHISM

INTRODUCTION

The distribution of metamorphic mineral assemblages within each sub-member of the Biwabik iron-formation has been described in the discussion of stratigraphy. In this chapter the paragenetic sequences shown by these assemblages will be summarized and their probable mode of origin discussed. It seems clear that the first stages of the general metamorphic event accompanying the emplacement of the gabbro resulted in thermal metamorphism, marked largely by the local development of fayalite, some ferrohypersthene, and possibly most of the lime-silicates of the uppermost two units of the Biwabik formation. The later stages of the same metamorphic event were distinguished by metasomatic activity which involved the introduction of water, carbon dioxide, calcium, magnesium, and traces of sodium, phosphorus, sulfur, arsenic, and copper into the wallrock surrounding numerous pegmatic veins. Hedenbergite, some ferrohypersthene, apatite, various feldspars and sulfides, loellingite, calcite, stilpnomelane, and all the amphiboles of the district were formed during the metasomatic event. Evidence supporting the previous existence of "primary" silicates and carbonates is absent in that all the metasomatic silicates listed above appear to have formed, in large part, directly from magnetite-quartz assemblages or silicates formed during the thermal metamorphic stage and not from "primary" silicates or carbonates. Furthermore, in those regions where the thermal metamorphic or metasomatic silicates have failed to develop, there are no relics, pseudomorphs, or inherited fabrics that would indicate in any way the earlier existence of such "primary" minerals but instead only relics of sedimentary structures displayed by fine-grained magnetite (locally with traces of late-stage martite) and quartz.

METAMORPHIC MINERAL ASSEMBLAGES

Most commonly the metamorphic silicates do not occur singly in a given specimen but are generally associated with other, nearly contemporaneous, silicates. Although cummingtonite is found in this manner, it is usually most common and abundant in taconite consisting essentially of quartz and magnetite. As stressed in the chapters on stratigraphy and mineralogy, each of the metamorphic silicates bears a consistent relationship with respect to the others. The relative ages shown among the most abundant metamorphic silicates are as follows: *

* Recent studies show that the occurrence of cummingtonite, anthophyllite, and dolomite assemblages near Mesaba records an even lower grade metamorphic zone lying between magnetite and quartz taconites containing only cummingtonite to the east, and only minne-

(oldest)   granoblastic fayalite, some prismatic-poikilitic ferrohypersthene
prismatic hedenbergite, some prismatic-poikilitic ferrohypersthene and possibly poikilitic cummingtonite
prismatic hornblende and actinolite
(youngest)   prismatic to acicular cummingtonite

Where these metamorphic minerals occur together, any one of the group can be poikilitically enclosed or partly replaced by any of the younger, and generally more hydrous, minerals below it on the list. Yoder (1957, p. 233) has pointed out that the various ferrous silicates can form upon metamorphism in a rock initially consisting of magnetite and quartz. An overwhelming amount of petrographic evidence indicates that this was the dominant process acting during metamorphism of the Biwabik formation, as discussed below.

It is significant that, without exception, the observed mineral paragenetic sequences are retrograde and that the mutual relationships among all the ferrous silicates are identical wherever they are found in any of the iron-formation submembers. On the basis of these data alone, however, it might not be valid to define isograd reactions or explain the retrograde metamorphic sequence within the entire formation, except in the general terms already stated. The greatest difficulty in this problem is that during the general metamorphic event a given mineral might conceivably have formed or not, depending upon locally differing conditions of metasomatism or because of initial differences in the composition of the sediments. A series of metamorphic isograds or a specific paragenetic sequence might not be justifiably established, for example, from separate occurrences of fayalite, ferrohypersthene, and cummingtonite assemblages found in three different stratigraphic units, but perhaps the proposed mineral sequence can be verified if the evidence can be supported in every detail from every other occurrence of these assemblages, which was true for the observed retrograde sequence. The definition and mapping of isograds in the metamorphosed iron formation is difficult because of the superposition of metasomatic activity upon the effects of an earlier thermal stage. Although most of the submembers contained cummingtonite whose abundance and occurrence varied laterally, many of the units were devoid of other silicates, and cummingtonite was found apparently having formed directly from quartz and magnetite throughout the area studied. Isograd zones based on abundance of minerals were crudely parallel in most units but were generally too broad and overlapping to be carefully delineated on a map. To approach this problem ideally, it would seem necessary to select a series of samples from a specific stratigraphic unit that contained a variety of silicates whose assemblages showed pronounced lateral variations. These ideal conditions are

sotaite or minnesotaite-ankerite assemblages to the west. The details of this zoning will be discussed at a future date.

nearly impossible to achieve, but by studying the mineralogical assemblages near an easily recognizable stratigraphic horizon the geologist can perhaps come the closest to choosing a suite of specimens of the same initial composition. It is assumed that the observed mineralogical assemblages would then depend largely upon the subsequent metamorphic and metasomatic conditions which have been imposed upon a suite of specimens of initially similar constituents. If regions corresponding to the occurrence or nonoccurrence of specific minerals, and if the reactions between these minerals can be stated, then specific isograd zones can be defined just as they might be in the description of the progressive metamorphism of a shale into a biotite-quartz gneiss. On the Eastern Mesabi, however, the over-all metamorphic sequence is retrogressive, and although specific mineral reactions can be specified, these reactions are found to occur nearly throughout the area studied and isograd zones thus defined lose much of their meaning. Although there are no differences in kind of reaction, there do exist differences in the lateral distribution and abundance of the main ferrous silicates found in the district. Thus, by taking the field approach mentioned, one can at least crudely delineate zones on the basis of the relative abundances of the "critical index minerals," which in this district are the ferrous silicates, and from their mutual stability relations specify a valid paragenetic sequence among them.

With this purpose in mind, specimens from above and below the contact between the Upper Cherty and Lower Slaty members of the Biwabik formation were collected for such an investigation.

PARAGENETIC SEQUENCE OF METAMORPHIC MINERALS

The majority of the specimens were collected from both the Upper Cherty and Lower Slaty members within 3 to 6 feet of their contact, although there were a few collected from as far as 10 feet or as close as one foot. These stratigraphic horizons were selected for detailed study because (1) the Upper Cherty–Lower Slaty contact is usually sharp and easily determined and is encountered in most of the holes in the area, and (2) other submembers do not contain as varied or as abundant assemblages of silicates and are thereby less suited for such a study. Except for some metasomatic silicates, particularly the calcium-rich varieties, the ferrous silicate assemblages of the specimens from the bottom of the relatively magnetite-rich Upper Cherty member (unit O) are essentially the same as those in the upper part of the relatively silicate-rich Lower Slaty member (unit P) from the same hole. The mineralogical differences between the pairs of specimens are in abundance rather than in kind. Identical mutual relationships were found to exist among the minerals of both members, and consequently the regional variations of the mineralogical assemblages present in submember P are very similar to the variations found in unit O.

The writers assume that when a given silicate is found occurring in different submembers, it has had an identical mode of origin in each of the submembers, because it is impossible to distinguish petrographically whether a given mineral, for example fayalite, has formed under "water-rich, high temperature" conditions, or in a "water-deficient, low temperature" environment. In the absence of evidence to the contrary from any of the other submembers, it is assumed that each individual of a given kind of ferrous silicate within the entire iron formation had an origin similar or identical to those proposed below for the Lower Slaty member, because the mutual relations among the ferrous silicates are identical to those of every other submember in which they occur.

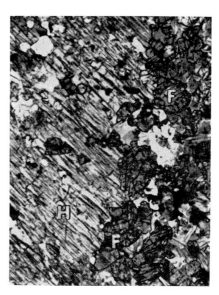

FIGURE 32. — Fayalite and ferrohypersthene. A typical occurrence of coarse poikilitic metacrysts of ferrohypersthene (light gray) enclosing fayalite (dark gray) is illustrated. Both the olivine and pyroxene are commonly enclosed, and in a few instances are partly replaced, by hornblende+cummingtonite and traces of hornblende. (15-49; 10X)

Specimens of unit P from holes 4 to 6, 8 to 11, and 13 to 17 are similar petrographically and can be described collectively as probably representing taconite closest to the gabbro during thermal metamorphism. These specimens all contain medium- to fine-grained fayalite, which invariably appears in a granoblastic mosaic with quartz and some magnetite. In some specimens, especially those from holes 4, 6, 10, 15, and 16, relatively coarser-grained poikilitic ferrohypersthene contains inclusions of fayalite, magnetite, and quartz. Trace to minor amounts of prismatic to poikilitic cummingtonite locally appear within the granoblastic mosaic and in most places appear to be contemporaneous or slightly later than the other silicates. In many specimens the quartz contains minute inclusions of, or is associated with, subhedral prisms of hornblende and hornblende+cummingtonite; in some places where these calcium-bearing silicates are locally abundant they commonly contain minute pleochroic haloes. The typical textural relationships between these minerals are shown in Figure 32, which shows the development of a ferrohypersthene metacryst within a granoblastic mosaic of fayalite and quartz. The matrix contains some hornblende+cummingtonite and minor amounts of hornblende that poikilitically enclose fayalite, quartz, and ferrohypersthene to a minor extent. Elsewhere in this slide, oligoclase accompanies quartz and hornblende. The relic lamellae of the specimen are defined by disseminated fine-grained magnetite. Only traces of fine-grained acicular cummingtonite have developed in these cores, and where found it has generally formed around grain boundaries of earlier minerals, or has clearly replaced the earlier fayalite, poikilitic ferrohypersthene, poikilitic cummingtonite, and quartz.

FIGURE 33. — Granule-shaped structures in fayalite. Clusters of fayalite metacrysts (gray) contain many granule-shaped outlines (dark gray) defined by extremely fine-grained magnetite, and possibly some graphite. A few recrystallized magnetite-rich granules (black) are also present. The remaining parts of the rock contain numerous slightly corroded remnants of fayalite in a granoblastic mosaic of fine-grained quartz and cummingtonite (white). (22-211; 12X)

Specimens from holes 18 to 22, 24, and 25 consist, in part, of mineral assemblages somewhat similar to those to the east except that these assemblages appear only as remnants within a quartzose matrix containing increasing amounts of fine-grained cummingtonite. Fayalite and quartz are present as granoblastic grains, some of which are locally enclosed by poikilitic prisms of cummingtonite. Poikilitic ferrohypersthene was present only in core from hole 21. Assemblages of these silicates occur as remnants within a matrix of granoblastic quartz and fine-grained cummingtonite prisms. The fayalite-bearing assemblages in many places have "wormy" or corroded borders and can be seen in many stages of replacement by the fine-grained cummingtonite. Many granule-like structures, partly composed of quartz, magnetite, and fayalite occur in these specimens. The granule-shaped clusters of fine-grained magnetite within the fayalite-rich parts are interpreted to be relic shapes of granules. This fabric is shown in Figure 33. Examples of the moderate to nearly complete replacement by fine-grained cummingtonite of granule-shaped fayalite grains, relic chert granules, and quartzose matrix are plentiful elsewhere in this group of specimens. The fayalite content of the specimens from these holes decreases from east to west and conversely the abundance of fine-grained cummingtonite increases. Almost without exception the fine-grained cummingtonite is intimately associated with minor to moderate amounts of hornblende+cummingtonite or minor amounts of hornblende. All three minerals locally contain small pleochroic haloes around minute radioactive minerals. In a specimen from hole 24, the fine-grained cummingtonite is accompanied by oligoclase and in hole 18 by apatite and brown (and minor green) biotite.

Specimens from holes 27 to 30, 32, and 33 contain small remnants of fayalite which are almost completely replaced by fine-grained cummingtonite. Probably fayalite was never extensively developed in this region, although there are many small remnants of fayalite within the extensively developed fine-grained cummingtonite. Many relic quartzose granules are partly replaced by fine-grained prismatic-acicular cummingtonite, as shown in Figure 34. A few remnants of fayalite also appear in this

FIGURE 34. — Fayalite remnants in cummingtonite. The replacement nature of cummingtonite with respect to fayalite is illustrated in this figure. Here a remnant of fayalite (lower left) is only partly replaced by prismatic cummingtonite (light gray). Traces of a serpentine-like mineral, probably hisingerite, formed along the replacement front between the above silicates. The granule-shaped clusters of cummingtonite usually contain only minute traces of the earlier fayalite. The matrix to these silicate-rich clusters consists of a decussate mat of acicular cummingtonite (medium gray) and xenoblastic quartz (white). (33-213; 30X)

photograph. Elsewhere in the same slide, where presumably fayalite has been completely replaced by fine-grained cummingtonite, the excess iron from the conversion has precipitated as fine-grained magnetite that apparently still preserves granule-like outlines. This specimen contains the westernmost remnants of fayalite of the submember, occurring about 8 miles along the strike of the formation from the east end of the Mitchell pit, but it is probable that the gabbro lies within 2 miles to the south. Because of its sill-like structure, the gabbro doubtless projected over the iron formation, and was probably closer to this specimen during metamorphism.

Fayalite is absent in specimens from holes 31, 36, and 37. Very fine-grained acicular cummingtonite and chert-like quartz dominates the constituents of the core. A few granule-like concentrations of disseminated fine-grained magnetite locally occur within the cummingtonite (Fig. 35), but it is impossible to prove petrographically that fayalite was ever abundantly developed in these silicate-rich specimens because there are no relics of any previous metamorphic or "primary" silicate. The abundant granule-like concentrations of fine-grained magnetite within the acicular cummingtonite of these holes, however, are practically identical to those found in cummingtonite that has nearly completely replaced fayalite in specimens from the next group of holes to the east.

Throughout the area, hedenbergite is found that has partly replaced or poikilitically enclosed fayalite; the pyroxene in turn has been partly replaced by differing amounts of cummingtonite, depending on the abundance of the latter in the given area. As discussed in the chapter on stratigraphy, the abundance and occurrence of hedenbergite are dependent solely on proximity to pegmatitic veins, and they show no zonal distribution with respect to the position of the gabbro. Many of the specimens from all the groups of holes have been cut by hisingerite veinlets and some of the adjacent silicates have been partly altered to green and brown hisingerite or brown nontronite. Clearly later than any of the metamorphic silicates which they cut, these veinlets are probably somewhat more abundant in the regions nearer the gabbro than in those farther away.

The foregoing descriptive petrographic information constitutes perhaps

FIGURE 35. — Granule-shaped structures in cummingtonite. Distinct relics of granule structures (dark gray) are common even in rocks that have been completely reconstituted. This specimen has been recrystallized in the presence of water and yet numerous granule-shaped clusters of very fine-grained magnetite clearly reflect the original granule fabric of the rock, which now consists almost entirely of prismatic and acicular cummingtonite.

the minimum amount of basic data necessary to propose or verify any metamorphic paragenetic sequence. Although the metamorphic mineral paragenesis as determined in other submembers is identical in every detail, it is believed that the paragenetic sequence resulting from metamorphism of a given petrologic system can be discussed and interpreted with confidence only if this field approach is used. Before attempting to explain this field data, perhaps it will be instructive to review some basic essentials of the simple petrologic system, $FeO-SiO_2-H_2O$.

THE IRON OXIDE-SILICA-WATER SYSTEM

Yoder (1957, p. 232) and Flaschen and Osborn (1957, p. 923) have discussed the implications of metamorphism of mineral systems containing iron oxides, silica, and water as components. A series of conventional tetrahedral and ternary diagrams for systems containing these components and the minerals observed in the Eastern Mesabi district is discussed below.

It is believed that prior to and during thermal metamorphism, almost all rocks of the Biwabik iron-formation consisted of constituents which lie within the quaternary system $Fe-Si-O_2-H_2$ as represented graphically in Figure 36, diagram a. The obvious exceptions to this generalization are the high carbonate (calcite) unit A; the possible local presence of aluminum in units which yield lime-silicates during thermal metamorphism; the carbonaceous material in unit Q; and the possible initial magnesium and manganese content of unit P of the Lower Slaty member and the bottom units M, N, and O of the Upper Cherty member. Subsequent to the thermal recrystallization, most of the submembers were locally metasomatized and the final discussion of the simple system mentioned above must take into consideration the widespread addition of water, magnesium, and to a lesser extent, carbon dioxide and calcium.

In addition to quartz and magnetite, the thermally metamorphosed Biwabik iron-formation also contained fayalite, probably some ferrohypersthene, and possibly traces of poikilitic cummingtonite; the components of all these mineralogical assemblages were essentially $FeO-SiO_2-H_2O-Fe_3O_4$. Consequently, the compositions of the minerals lie within the volume defined by the tetrahedron shown in Figure 36, b.

It has been observed that magnetite, and not hematite, is the iron oxide that appears to be in stable equilibrium with the iron silicates of this petrologic system. Chemical analyses of the dominant iron-bearing silicates of both thermal and metasomatic origin show that they are all essentially ferrous silicates. These observed conditions most closely correspond to the experimental investigation of the system iron oxide-silica-water at low partial oxygen pressures carried out by Flaschen and Osborn (1957, p. 935). Under these conditions the stable crystalline phases are the ferrous silicates and magnetite. These authors also state (p. 930) that in this environment, in which the iron-silicates are essentially ferrous, the compositions of the stable crystalline silicate phases lie close to the FeO–

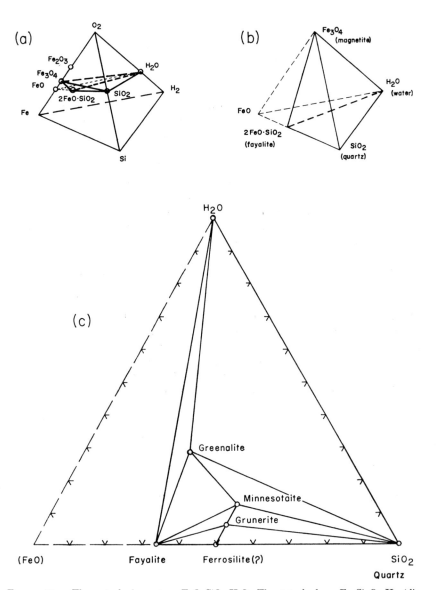

FIGURE 36. — The petrologic system FeO–SiO$_2$–H$_2$O. The tetrahedron Fe–Si–O$_2$–H$_2$ (diagram a) shows the location of the system fayalite-quartz-water-magnetite (diagram b). The projection of the system fayalite-quartz-water-magnetite onto the FeO–SiO$_2$–H$_2$O plane at approximately room temperature is presented in diagram c. The composition FeO is represented by iron + magnetite at this temperature. (Yoder, 1957.)

SiO_2–H_2O plane, shown in Figure 36, c (after Yoder, 1957). Unfortunately, these authors were unable to obtain grunerite experimentally as a stable phase. It should be further pointed out that the work of Flaschen and Osborn was not confirmed by Smith (1957, p. 230).

Yoder (1957, p. 233) has plotted the compositions of the ferrous silicates in the FeO–SiO_2–H_2O plane and supplied the joins that are believed to exist in equilibrium with magnetite at room temperature, as shown in Figure 36, c, and proposes (p. 232): "All the possible phases are present at the lowest temperatures, yet only those assemblages greenalite+quartz+water (analogous to a sediment) or greenalite+fayalite+water (analogous to a partly serpentinized dunite) are commonly observed. In the sediments, for example, the phases are in equilibrium with water, and, therefore, only those assemblages in which water can occur as a phase are permissible. The remaining assemblages may occur in environments where water does not exist as a phase." In this system, Yoder continues, water "represents the homogeneous gas phase, the composition of which, although close to H_2O, may be enriched in the component oxygen or the component hydrogen as well as iron and silica. If the component oxygen is in excess of that of the saturated gas in equilibrium with magnetite and silicates, all the silicates and some or all of the magnetite would be oxidized to an assemblage of magnetite+hematite+quartz+gas or hematite+quartz+gas."

Flaschen and Osborn (1957, p. 935) state that in the environment of low partial oxygen pressure where the iron-silicates are essentially ferrous —which closely agrees with the conditions of metamorphism in the Eastern Mesabi district—the water phase is represented by the H_2O apex of the plane but actually has a higher hydrogen:oxygen ratio than is indicated by H_2O and it contains iron oxide and silica in solution.

Both Yoder, and Flaschen and Osborn have considered the possible variations of mineral assemblages that might accompany increasing temperatures of progressive thermal metamorphism. The total composition of a given assemblage is fixed by a given FeO:SiO_2:H_2O ratio and its total oxygen content remains essentially constant. It will be noted that the presence of hydrous, iron-rich "primary" silicates is implied in their discussions. Flaschen and Osborn (1957, p. 942) suggest the following succession of mineral assemblages: magnetite-greenalite (up to 250°C), greenalite-fayalite (250°–470°C), fayalite-minnesotaite (470°–480°C), and fayalite-quartz (above 480°C). On the other hand, Yoder (1957, p. 234) has deduced that a sediment with an initial bulk composition of greenalite-quartz-magnetite-water would yield minnesotaite, then grunerite, and finally fayalite or hypersthene with progressive metamorphism. He proposes that the appearance of each of these index minerals will be determined by the following reactions at successively elevated temperatures.

greenalite + quartz → minnesotaite + vapor
greenalite (decomp.) → fayalite + minnesotaite + vapor

minnesotaite + fayalite → grunerite + vapor
minnesotaite (decomp.) → grunerite + quartz + vapor
grunerite (decomp.) → fayalite + quartz + vapor

He also notes that irrespective of temperature, all assemblages are in equilibrium with magnetite, and that at the highest temperature the stable assemblage is magnetite-fayalite-quartz-vapor.

Because all of these iron-bearing silicates are presumed to be present in the Biwabik formation, it should be possible, therefore, to determine isograds (specifically defined by one of the above reactions) of progressive metamorphism which would be distinguished by the appearance of the index minerals, minnesotaite, grunerite, ferrosilite ("hypersthene"), and finally fayalite. Consequently, field studies and petrographic examinations should reveal that, laterally within any given submember, minnesotaite-bearing assemblages are found partly replaced by grunerite-bearing assemblages; and at another point these assemblages are found partly replaced by ferrosilite-bearing assemblages, which in turn have been reconstituted into fayalite-bearing assemblages as the regions of increasing thermal intensity are approached, which in this instance implies increasing proximity to the Duluth gabbro. One of the most important initial objectives of the entire regional study was, in fact, to take this exactly same approach in order to uniquely define the isograd zones and to specify the successive mineral reactions of progressive metamorphism. After examining miles of freshly exposed rocks in the Mitchell pit and studying hundreds of thin sections, the writers were forced to conclude that evidence of progressive thermal metamorphism of "primary" silicates into the silicates currently found in any unit throughout the district was entirely lacking. It was not until near the end of the second field season (1956), when the exposures in the Mitchell pit became sufficiently extensive, that the significant role of widespread metasomatic activity in forming the great bulk of silicates was clearly recognized. Significantly, Yoder (1957, p. 233, and personal communication) has made the point that the various iron silicates can, however, form under reducing conditions during metamorphism in beds initially consisting of magnetite and quartz; all field and petrographic evidence viewed by the writers fully supports this statement.

INTERPRETATION OF OBSERVED PETROLOGIC
SYSTEM $FeO-SiO_2-H_2O-MgO$

For the presentation and explanation of the petrographic data, only the relationships among the minerals lying in the $FeO-SiO_2-H_2O$ plane (Fig. 36, c) will be discussed. On the basis of the petrographic studies and chemical analyses of the ferrous silicates that lie within this plane, it must be understood that magnesium is also an important component of this petrologic system although it is not represented graphically. For simplicity, the actual system is perhaps best represented by Figure 37,

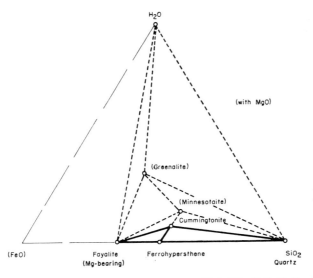

FIGURE 37. — Observed petrologic system FeO–SiO_2–H_2O–MgO.
(After Yoder, 1957.)

modified after Yoder (1957), which implies the presence of magnesium (either initially or metasomatically) in amounts sufficient to form the observed minerals. The dashed joins represent assemblages that might occur at lower temperature in the presence of sufficient water phase, but which were not observed in the area studied.

Utilization of such a graphic aid as the FeO–SiO_2–H_2O plane necessitates definition of the system to which it is being applied. A point of the plane may represent the bulk composition of a cubic foot of the rock, or that of a group of lamellae, or even that of a single granule, provided that free interchange of constituents took place during metamorphism. Near the gabbro and pegmatites, prolonged high temperatures probably permitted substantial interchange of constituents. Farther from the gabbro, abundant aqueous fluids at lower temperatures aided interchange and reconstitution of material as evidenced by abundant cummingtonite and other amphiboles.

The writers have proposed that a given granule or lamella, prior to metamorphism, consisted mainly of variable amounts of iron oxide (magnetite) and SiO_2 and possibly trace amounts of magnesium and aluminum. On the basis of an examination of thin sections of various submembers (particularly of P), which he viewed during a discussion with the writers, Yoder has agreed that it is possible that the granules, prior to metamorphism, were of the proposed composition. The granules and lamellae that are preserved as relics indicate these small-scale structures or portions of the iron formation once varied in composition from nearly pure chert to nearly pure magnetite, and that in some places there were sufficient amounts of "impurities" to permit the formation of, for example, magnesium- and manganese-bearing fayalite. Although the granules were apparently of chemical origin, they were undoubtedly transported, sorted, broken, packed together, and generally acted upon by normal transportation agencies prior to final deposition. This mixing, along with an inherent variability of initial composition dependent upon the lithotope in which the granule first grew, easily explains the compositional variation among the adjacent granules as well as the matrix which encloses them.

In order to explain the observed paragenesis, it is proposed that the first phase of the general metamorphic event was thermal, induced by the emplacement of the Duluth gabbro. During the thermal stage some of the original iron and silica constituents of the iron formation probably existed as unstable components in the presence of traces of magnesium, aluminum, and possibly manganese. It is clear, of course, that in any event the partial pressure of oxygen must have been sufficiently low to permit the reconstitution of the components into ferrous silicates. These components ultimately reacted and were reconstituted to yield magnesium-manganese-bearing fayalite, magnetite, and quartz, locally with poikilitic ferrohypersthene as stable crystalline phases, as indicated in the studies of Yoder and of Flaschen and Osborn. Presumably all the observed fayalite-quartz-magnetite assemblages, with or without ferrohypersthene, formed during this stage. During the incipient cooling stages, traces of poikilitic cummingtonite were permitted to form where water was available and cool enough to enter the crystalline phases. There are two possible explanations for the relative absence of hydroxyl-bearing silicates in these assemblages, but it is impossible to determine which actually occurred. In the first case, the observed fayalite-bearing assemblages might represent metamorphosed sediments that were initially water-deficient, with the implication that the metamorphism could have taken place at low temperature; possibly as low as 250°C if there were insufficient water to produce an aqueous phase (Flaschen and Osborn, 1957, p. 940). Although it is believed that the iron formation was water-deficient prior to metamorphism, the temperature of metamorphism was certainly much higher. Because of the great abundance of retrograde, fine-grained cummingtonite that formed during the waning metasomatic stage, the writers believe that the fayalite along with some poikilitic ferrohypersthene and poikilitic cummingtonite assemblages formed under conditions of initial high temperature, probably on the order of 600°C to 700°C. The second alternative is that water might indeed have been present as a phase, but at the high temperatures of the thermal stage little could enter the stable hydrous crystalline phases of the system to form amphiboles, such as grunerite. By the time the regions near the gabbro had cooled sufficiently to form stable hydrous silicates, it is likely that nearly all of the water phase had migrated into the cooler regions far from the gabbro. The presence of some water as a vapor phase probably facilitated the diffusion of material within the rock mass and thereby permitted the growth of coarse crystalline phases, but it is believed that the sediments were essentially water deficient prior to metamorphism.

Perhaps the most significant question to ask at this point is, Why was the formation of abundant fayalite restricted essentially to one stratigraphic interval, generally about 100 feet thick, within the lower parts of the formation throughout the district? Of course one conclusion to draw is that this is a stratigraphic control governed solely by initial composition. If this were the case, then, it is necessary to assume that in other parts of the formation the necessary constituents and possibly other "catalytic impurities" were either entirely absent or not present in sufficient amounts to bring about the recombination of quartz and magnetite constituents into fayalite-bearing assemblages regardless of high temperatures. It must be remembered that, at every point in the district, the main fayalite-bearing zone of the Lower Slaty and Upper Cherty members is overlain by approximately 200 stratigraphic feet of iron formation which was nearer the gabbro and therefore significantly hotter during thermal metamorphism than the zone which actually yielded abundant fayalite. Except for a very few local traces, particularly in the Lower Cherty member, fayalite is virtually absent elsewhere in the iron formation. Many parts of the iron formation are now richer in magnesium, by virtue of the locally abundant cummingtonite, but this element is metasomatic in origin and reflects little if at all upon the initial composition of the sediments. Although sufficient thermal energy was certainly available, and there was undoubtedly sufficient time for the reconstitution of quartz and magnetite into fayalite, the latter is still virtually absent in the Upper Slaty member and the upper two-thirds of the Upper Cherty member. Consequently, it is necessary to conclude that although the necessary components, in quartz and magnetite, and thermal energy for the formation of fayalite were probably available throughout the iron formation, only one thin zone had a sufficiently reducing environment so that the partial pressure of oxygen was low enough to permit the reconstitution.

The obvious conclusion reached above now gives rise to another question as to the localization of the reducing environment during thermal metamorphism. In the discussion of stratigraphy (Chapter 2), it was pointed out that unit Q is generally rich in graphitic material and that unit P is marked by thin graphitic partings and lamellae in many places. Chemical tests by James Palacas have shown the presence of traces of organic material in unit Q, and it is more than likely that the graphitic material throughout the Lower Slaty was derived from organic carbonaceous material. During thermal metamorphism, it is probable that the presence of the organic material was instrumental in producing the local reducing environment necessary for the formation of abundant fayalite and some ferrohypersthene in units P and Q. Because graphitic material is rare in the overlying units M, N, and O, as well as in the underlying units R, S, T, and V, it is necessary to assume that the presence of escaping organic material from the adjacent units P and Q, collectively about 85 feet thick on the average, locally produced reducing environments sufficient to permit the formation of fayalite and ferrohypersthene next to the Lower Slaty member. The writers are not aware of any other alternative reason for the sufficiently low partial pressure of oxygen to yield fayalite within such a locally restricted stratigraphic zone. This interpretation would explain the preservation intact, save for local recrystallization, of the relatively simple quartz-magnetite assemblages in relic granules and lamellae which are so abundant in the remaining iron-rich quartzose members that contain little or no fayalite.

The next step in the observed paragenetic sequence is a metasomatic stage which is almost contemporaneous with, but which probably shortly followed, the formation of the fayalite-bearing assemblages. Extensive pit exposures allowed detailed study of the metasomatic nature of the minerals in the wallrock adjacent to numerous pegmatites. During emplacement of the pegmatitic veins, calcium, magnesium, water, carbon dioxide, and lesser amounts of sodium, phosphorus, sulfur, arsenic, copper, molybdenum, and aluminum were introduced into the iron formation. The mineralogy of the pegmatites and the adjacent wallrock has been discussed in Chapter 1. In brief review, the discordant veins consist of widely differing amounts of quartz, alkali feldspar (albite), silicic plagioclase (oligoclase), hornblende, green biotite and secondary chlorite, apatite, pyrrhotite, magnetite, and stilpnomelane. Ferrohypersthene, andradite, muscovite, loellingite, chalcopyrite and secondary covellite, molybdenite, pyrite, and hedenbergite are present in some pegmatites but are not typical of the assemblages commonly found. Only the ferrous silicates will be considered in the discussion below.

Two pyroxenes, hedenbergite and ferrohypersthene, are commonly found in the wallrock immediately adjacent to the relatively large and mafic veins. Generally, the most abundant and obvious is hedenbergite, which developed most commonly in taconites consisting essentially of quartz and magnetite; although these rocks were usually recrystallized, wholly to partly replaced relic magnetite-quartz beds and granule structures are often found preserved. Where the pegmatites have been emplaced in fayalite-bearing taconites, hedenbergite always poikilitically encloses or has substantially replaced the fayalite. The distribution of some medium- to fine-grained ferrohypersthene is clearly related to the thermal metamorphism accompanying the emplacement of the gabbro, as discussed above. A second occurrence of ferrohypersthene is locally typical of the wallrock adjacent to some of the relatively mafic pegmatite veins. In this occurrence the pyroxene typically consists of slightly poikilitic, extremely coarse- to coarse-grained crystals that have formed in clusters which diminish in size and abundance away from the veins. Occurrences of this variety of ferrohypersthene, formed directly from quartz and magnetite, are rare because the pyroxene is almost invariably found as large remnants in a matrix of actinolite, and hornblende+cummingtonite or cummingtonite. This occurrence of the pyroxene is clearly related to pegmatitic activity, and consequently is of about the same relative age as hedenbergite, but it is impossible to determine the amounts of magnesium initially present in the wallrock or the amount metasomatically introduced.

Examinations of the pit exposures revealed taconite consisting almost entirely of fine- to medium-grained quartz and magnetite, normally occurring adjacent to the areas of extensive metasomatic effects, for example, outward beyond the zones of abundant development of hedenbergite associated with the most mafic veins. Although the mineral assemblages in the adjacent area are essentially anhydrous next to the largest veins, it is probably necessary to postulate the presence of a water phase as a medium of transport for the wide dispersal of the hedenbergite. It seems clear that during the initial stages of metasomatism the temperature was too great to permit water to enter any stable crystalline phases.

Eventually the temperature of the taconites adjacent to the veins entered the range within which hydroxyl-bearing silicates, particularly amphiboles, joined the metamorphic mineral assemblages of stable crystalline phases, as implied in Figure 37. In the areas where the mafic pegmatites are thick, and are presumed to have heated the wallrock to a greater extent than could the thinner veins, the magnesium-rich metasomatic amphiboles of cummingtonite formed in the outermost zone beyond the anhydrous zone immediately surrounding the pegmatites. This is the most common mode of occurrence of cummingtonite. The amphibole evidently formed directly from quartz and magnetite throughout the remainder of the formation, and invariably it has incipiently to nearly completely replaced fayalite wherever the two occur together in a given assemblage. These observations, however, merely date the amphibole as post-olivine. In many places, particularly next to the thinner mafic veins, cummingtonite has developed within the anhydrous zone, and here it has incipiently to almost completely replaced hedenbergite, ferrohypersthene as well as fayalite, and quartz and magnetite wherever any of these mineral assemblages have formed in the rock adjacent to the veins. In many places where sufficient quantities of metasomatic calcium and aluminum were also introduced in addition to magnesium and water, the amphiboles hornblende, actinolite, and hornblende+cummingtonite have a similar mode of origin and show identical replacement relationships. It has been pointed out in an earlier section that the various ferrous silicates can form during metamorphism directly from quartz and magnetite under certain conditions.

Referring to Figure 36, c, and Figure 37, it is clear that any of the hydrous silicates can be obtained from fayalite and (or) ferrohypersthene (or ferrosilite) by the addition of water to the system without appreciably changing the temperature or pressure. The same results can also be achieved by decreasing the temperature at constant water content for the system, but the writers believe that the former alternative best accounts for the retrograde assemblages found in the Eastern Mesabi district. Although the composition of hedenbergite, hornblende, and actinolite cannot be easily represented in the compositional systems discussed above, it is not difficult to envision the additional presence of calcium and aluminum in the metasomatizing fluids. It should also be pointed out that stilpnomelane is commonly found as an additional retrograde product of calcium-rich amphiboles both in the pegmatites and the iron formation. Apparently, some of the late-stage green and brown biotite, chlorite, and possibly traces of hisingerite that are more closely associated with the development of cummingtonite are of the same relative age as the stilpnomelane found with the calcic amphiboles and quartz-calcite-stilpnomelane veinlets.

The relationship between the fine-grained cummingtonite and the fine-grained minnesotaite of the Mesabi range to the west is not known at present, but it is believed that minnesotaite is a mineral of metamorphic origin. Although specimens from the transitional area contain only one mineral or the other, more recent preliminary work has shown that anthophyllite occurring with cummingtonite-dolomite assemblages is of younger age than the latter amphibole. Specimens showing the relationship between anthophyllite and minnesotaite have not been found.

THE "PRIMARY" SILICATES OF THE BIWABIK FORMATION

In the light of the earlier studies of the Mesabi range (especially Leith, 1903, p. 239 and Gruner, 1946, p. 12), the possibility that the initial iron formation consisted largely of thin layers and granules of primary silicates, carbonates, and iron oxide particles within a cherty matrix was proposed. The granule nature of the cherty layers was believed to be a primary sedimentary fabric. The granule structures were assumed to have initially consisted of various combinations of siderite, greenalite, minnesotaite, and stilpnomelane, and were assumed to be chemical precipitates. In the opinion of the writers, however, the hydrous silicates are not of primary, diagenetic, or authigenic origin, but formed at a later date under conditions of low temperature hydrothermal metamorphism. There are no relics of the "primary" minerals or pseudomorphs that show any evidence of their former existence in the Eastern Mesabi district, and all carbonates in the district are calcium-bearing and clearly of metasomatic origin. For the purposes of discussion, however, let us assume for the moment that the granules and some of the lamellae of the iron formation were composed, at least in part, of the "primary" silicates prior to any metamorphic activity.

If this were the case, it is now necessary to account for the reconstitution of the "primary" silicates greenalite, minnesotaite, and stilpnomelane (and possibly others) within a given structure — for example, a granule — directly into quartz and magnetite without the formation of intermediate amphiboles or other silicates, while still excellently preserving intact the delicate granule structure. It is also necessary to infer that the process must have taken place in almost all submembers of the formation throughout the entire area to produce the distribution of quartz and magnetite granules actually observed.

The experimental work of Flaschen and Osborn (1957, p. 937) shows that greenalite decomposes to fayalite, minnesotaite, and water at about 470°C and that minnesotaite decomposes to fayalite, quartz, and water at about 10°C higher. Yoder (1957, p. 234) has also stated that the assemblage stable at highest temperatures is fayalite, quartz, magnetite, and water. Gruner (1946, p. 64) found that greenalite heated for three days at 650°C in an atmosphere of CO_2 was entirely converted to fayalite. None of these observers mentions the formation of quartz and magnetite alone as products of thermal decomposition of the primary silicates. Yoder believes (personal cummunication), as do the writers, that under conditions of thermal metamorphism the breakdown of greenalite, minnesotaite, or stilpnomelane to quartz and magnetite, without the formation of amphiboles or other silicates, is unlikely.

In the consideration of the thermal decomposition of the primary silicates we have actually assumed that the total oxygen content is constant and that the oxygen partial pressure is low. Flaschen and Osborn (1957, p. 942) have considered the possibilities of changing the mineralogical assemblages by varying the oxygen partial pressure as well as the temperature during metamorphism. Their experimental work indicates that an assemblage defined by a given $Fe:SiO_2:H_2O$ ratio under very low partial pressure of oxygen may progressively change, for example, through the following sequence merely by increasing the partial pressure of oxygen at constant temperature:

magnetite + greenalite
magnetite + greenalite + minnesotaite
magnetite + minnesotaite
magnetite + minnesotaite + quartz
magnetite + quartz

where each assemblage is in equilibrium with an aqueous phase. Each successive stage represents an increase in the oxygen content of the system. This process may be described pictorially by referring to Figure 36, a. Compositionally speaking, the ferrous silicates lie in the $FeO:SiO_2:H_2O$ plane, and any primary magnetite-quartz granules or lamellae correspondingly lie in the $Fe_3O_4:SiO_2:H_2O$ plane. Actually these "primary" silicates, although essentially ferrous, do have varying $Fe_2O_3:FeO$ ratios and thereby lie within the volume between these planes. With the addition of oxygen it is possible that magnetite and quartz might completely replace the silicates if the $Fe_2O_3:FeO$ ratio increases to one (Flaschen and Osborn, 1957, p. 940). It is therefore possible that some of the ubiquitous magnetite-quartz granules and lamellae were formed by the oxidation of the silicates which they originally comprised. It is highly improbable, however, that all the different "primary" silicates, with initially differing ferric:ferrous ratios, can be oxidized exactly to the $Fe_3O_4:SiO_2$:-H_2O plane of Figure 36, a, and not beyond (lest hematite join the premetamorphic mineral assemblage), and that these silicates can be selectively oxidized in a rock also containing magnetite and quartz without the oxidation of appreciable amounts of magnetite to hematite. The remote possibility nevertheless exists.

As an alternative, the entire problem may be dismissed by assuming a highly extraordinary distribution of the initial constituents in which only parts of granules, lamellae, etc. (as well as only parts of submembers), which now consist of metamorphic silicates, originally consisted of the "primary" silicates with the remainder consisting of quartz and magnetite. This extraordinary initial distribution would be remarkably fortuitous with respect to the proximity of these constituents to the present gabbro mass and to the pegmatitic veins that cut the formation. It would also be exceptionally remarkable that in every instance the gabbro and pegmatites were just sufficiently hot to reconstitute all "primary" silicates without leaving any remnants along the outermost zones of metamorphic silicate formation. Another proposal might be that the entire Eastern Mesabi sequence of iron-formation rocks is different because these rocks were deposited within the "oxide facies" of James (1954, p. 256), which certainly appears to have taken place, but the writers believe that this also occurred on the Mesabi range in general, except possibly for the occurrence of carbonate in the far west.

The writers are unaware of any other mechanism by which the primary silicates could have been transformed into quartz and magnetite while still preserving the primary sedimentary fabrics as relics, and which could be relied upon to substantiate the observation that quartz and magnetite appear to have constituted almost all of the iron formation in the Eastern Mesabi district prior to the emplacement of the Duluth gabbro and numerous pegmatites and prior to the formation of the ferrous metamorphic silicates.

In summary, the ubiquitous occurrence of unquestionably relic quartz-magnetite granules and lamellae in almost every submember of the iron formation, and the formation of cummingtonite, hornblende, actinolite, hornblende+cummingtonite, hedenbergite, ferrohypersthene, and fayalite directly from magnetite-quartz assemblages (with metasomatism where necessary)—as well as the complete absence of "primary" silicates or relics of their former existence where any of the metamorphic silicates have not developed—most strongly suggest that prior to metamorphism the composition of the iron formation in the district was more nearly represented by quartz and magnetite than by the so-called primary silicates. In conclusion, the writers can cite no evidence from the Eastern Mesabi district, or indeed the entire Mesabi range, that would invalidate the concept that the iron formation originally consisted largely of chert and magnetite and that any of the "primary" and "metamorphic" silicates and carbonates can be formed directly from these minerals under suitable conditions of temperature and available water and carbon dioxide.

SUMMARY AND CONCLUSIONS

The early stages in the study of the effects of metamorphism in the Eastern Mesabi district were spent in attempting to recognize isograd zones and define specific reactions involving critical index minerals. Many earlier workers believed that the pre-existing "primary" silicates were destroyed by metamorphism near the gabbro and reconstituted into iron amphibole, grunerite-cummingtonite, actinolite, fayalite, and chlorite (Grout and Broderick, 1919, p. 12; Gruner, 1946, p. 186; Richarz, 1927, p. 705). It was reasoned that relics of the "primary" minerals and initial sedimentary fabrics would necessarily be instrumental in deciphering the paragenetic sequence of the expected effects of progressive metamorphism. There are literally myriads of relics of primary sedimentary textures or structures, composed of quartz and magnetite, especially granules, poorly-layered chert beds (now largely quartz), intraformational conglomerate pebbles and thin lamellae, all excellently preserved in nearly every submember of the Biwabik formation. In many instances the relic granule fabric is only partly preserved or merely outlined within recrystallized or reconstituted metamorphic mineral assemblages, but the least destroyed, or most delicately preserved, of the relic sedimentary textures always consist of extremely fine-grained magnetite in very fine-grained quartz. Whether these unquestionably primary structures were initially composed of magnetite and quartz at the time of deposition or are the result of subsequent diagenesis is impossible to prove from petrographic evidence. The presence of a given mineral distributed within a delicately preserved fabric or structure, regardless of how fine-grained it is, is not conclusive proof that the material was primary. This argument cannot be used to prove that a particular very fine-grained granule or lamellae, now consisting of chert and magnetite or of greenalite, initially consisted of these constituents just because they are extremely fine-grained.

Numerous thin sections of rocks from throughout the entire district show the reconstitution of relic magnetite-quartz structures, in some cases with metasomatic additions, directly into fayalite, ferrohypersthene, hedenbergite, hornblende, actinolite, hornblende+cummingtonite; and cummingtonite; none of these silicates was found in the process of formation from "primary" silicates or carbonates. With the exception of carbonate (calcite) and carbonaceous beds, the earliest discernible mineral assemblages in the taconites are always quartz and magnetite, and where no metamorphic minerals have formed in these rocks, remnants or any evidence whatsoever for the former existence of the so-called primary silicates and carbonates are entirely lacking. The iron-bearing dolomites of the iron formation near Mesaba are clearly metasomatic and are genetically associated with cummingtonite and anthophyllite. In any event, all of these minerals are collectively found to have partly to wholly replaced relic sedimentary structures composed of magnetite-quartz assemblages. In summary, all the observational evidence makes it necessary to conclude that at least prior to metamorphism, the Biwabik iron-formation in the district consisted mainly of quartz and magnetite.

The results of field and petrographic examination revealed no evidence for progressive metamorphism of hydrous "primary" silicates into more anhydrous metamorphic silicates with increasing proximity to the Duluth gabbro. According to optical properties and chemical analyses, the important metamorphic silicates of the district are essentially ferrous, indicating metamorphic recombination of the iron-formation constituents took place in the presence of sufficiently low partial pressures of oxygen. Within every submember the paragenetic sequence of the ferrous silicates is identical; although, of course, not every metamorphic silicate occurs in each of the units of the formation. Wherever a pair or group of ferrous silicates is associated, however, without exception the given individuals always bear the same relationship to one another. In all cases, it was found that the paragenetic sequence during metamorphism is that of retrograde formation of more hydrous silicates from more anhydrous silicates.

It is proposed that the first phase of the over-all metamorphic event was thermal, resulting largely in simple recrystallization as well as the formation of new fayalite-magnetite-quartz assemblages — with and without subordinate to trace amounts of ferrohypersthene and poikilitic cummingtonite. The distribution and abundance of fayalite are clearly related to the proximity of the gabbro mass, although its occurrence is restricted almost entirely to one stratigraphic interval. Apparently, fayalite would form abundantly only in that part of the stratigraphic column containing carbonaceous material, the decomposition of which provided a sufficiently reducing environment (low partial pressure of oxygen) for the reconstitution of the constituents of quartz and magnetite. The remainder of the submembers at this stage in the general metamorphic event still consisted largely of locally recrystallized magnetite and quartz assemblages, and most of the primary sedimentary fabrics were only slightly modified during recrystallization.

Almost contemporaneously with, or shortly after the climax of, the thermal stage, a metasomatic stage was initiated. In addition to feldspars, biotites, apatite, sulfides, and loellingite, the most apparent minerals of metasomatic origin are hedenbergite, some ferrohypersthene, various calcium-rich amphiboles, and cummingtonite. On a regional scale the occurrence and abundance of the calcium-bearing silicates is seemingly erratic, but detailed field studies reveal that they are directly related to the proximity of pegmatite veins. All of the metasomatic silicates are found to have formed directly from quartz and magnetite as well as from fayalite by replacement; consequently, metasomatic reconstitution must have taken place under sufficiently reducing conditions during metamorphism. Cummingtonite is the most extensively developed metasomatic silicate, and although it replaces all the other anhydrous silicates of the district, it is most commonly found as having developed directly from magnetite and quartz. Cummingtonite, generally abundant in the district, has not commonly formed in the regions closest to the gabbro or to the pegmatites. It is probable that water introduced into the surrounding iron formation diffused from the local regions of highest temperature adjacent to the pegmatites outward into the cooler wallrock regions where stable hydrous silicates could form. This type of mineral zoning was observed surrounding the pegmatites in the Mitchell pit. Water introduced from pegmatites in the part of the formation near the grabbo, essentially the region immediately southeast of Argo Lake, did not result in the abundant formation of cummingtonite because during the introduction of water the temperature of the wallrock was probably too great to permit the formation of the hydrous silicates. By the time these rocks had cooled sufficiently, most of the water had migrated to more distant and cooler parts of the formation. This mechanism explains the minute amounts of retrograde cummingtonite in the cores and hand specimens from near the gabbro. Slightly farther away from the gabbro, the abundance of the amphibole increases and the earlier fayalite-bearing assemblages have been somewhat corroded and incipiently to partly replaced by fine-grained cummingtonite. The fact that fine-grained cummingtonite is most abundant in the area south and west of Iron Lake, where few remnants of fayalite persist in it, supports the conclusion that the water phase migrated away from the high temperature region near the gabbro.

5. PRACTICAL ASPECTS

GENERAL DISCUSSION

The rapid depletion of the enriched types of iron ores from the Mesabi district, especially during and after World War II, led to increasing interest in the enormous reserves of magnetite in the Biwabik iron-formation. The experimental work of the Mines Experiment Station at the University of Minnesota, under the direction of E. W. Davis, showed the possibility of producing a high-grade concentrate from the magnetite taconites. It remained to be demonstrated whether the process could be made commercially feasible. The Reserve Mining Company, later the Erie Mining Company, and the Oliver Iron Mining Division of the U.S. Steel Corporation undertook the task of proving the practicability of concentrating the magnetite from the exceedingly hard, fine-grained taconite.

With the beginning of large-scale mining and concentration it became evident that much more precise mineralogical and geological information was needed for the Eastern Mesabi district.

In a general way it was known that the nature of the Biwabik formation was somewhat different east of the town of Mesaba because there was an absence of secondary concentration which had made the rich orebodies of the district from Mesaba westward. This was rather obviously due to the nearness of the huge Duluth gabbro intrusion to that part of the Biwabik formation (Grout and Broderick, 1919).

It was also known that the nature of the formation changed vertically as a result of variation in sedimentation during deposition of several hundred feet of chemical sediments (Broderick, 1919).

As is often true, detailed work showed that the eastern part of the district, a length of about 18 miles, was much more complex than previously recognized. This, in turn, had a direct bearing on the problems arising in mining and concentration of the ore. The emplacement of the transgressing Duluth gabbro and that of numerous pegmatitic veins cutting the iron formation resulted in extensive thermal and metasomatic metamorphism. Its effect was to locally reconstitute the previously existing minerals of the Biwabik formation, in large part quartz and magnetite, into a variety of silicate mineral assemblages (Gundersen, 1958) which locally affect the recoverability of the remaining magnetite.

As previously discussed in this bulletin, the mineralogical variations within the magnetite-bearing taconites are commonly abrupt, both vertically and laterally. In order to indicate in general terms the influence of initial bedding textures and the effect of later development of metamorphic silicates upon the beneficiation characteristics of the magnetic taconites, two drill cores will be discussed in terms of their stratigraphy, taconite lithology, and magnetic "concentratability" as determined by Davis tube testing. Figures 38 and 39 show the simplified stratigraphy in terms of the Lower Cherty, Lower Slaty, Upper Cherty, and Upper Slaty (Wolff, 1917) members and the many submembers (Gundersen, 1958), which are designated by letters. Four curves are also plotted on these figures to indicate the results of magnetic tube tests corresponding to the stratigraphic units. Curve 1 shows the variation in weight per cent of iron occurring as magnetite in the crude taconite feed. Curve 2 indicates the weight per cent of total iron in the same feed. The "recovery" or weight per cent of original total iron recovered in the magnetic concentrate is indicated by curve 3. Curve 4 records the "grade" or weight per cent of iron in the magnetic concentrate of curve 3. The grade of concentrate, calculated on a basis of 85 per cent minus 325 mesh, is plotted as the solid portions of the curve; but where the total recovery is low, the actual grade at minus 150 mesh is plotted as the dashed portions of curve 4 because that was the only data available.

The data from the holes presented here indicate the total iron content (curve 2) of the iron formation ranges essentially from 25 to 35 weight per cent, which is usually average for the district. The separation between curves 1 and 2 of the diagrams indicates the amount of iron that is present in an unrecoverable form, mainly as iron-rich metamorphic silicates.

The potential recovery of iron from any unit, curve 3, is obviously directly dependent upon the amount of iron in a recoverable form, i.e., the amount of magnetite present (curve 1). The reasons for the variation of grade of magnetic concentrate, however, are less apparent in terms as simple as these. Consequently, the emphasis in this chapter is directed mainly at the influence of grain fabrics found in relic bedding structures of the various taconite types, as well as the local effects of new mineral assemblages of metamorphic and metasomatic activity, upon the potential grade of magnetic concentrate that might be recoverable from a given stratigraphic unit.

A line marking a "marginal" 64 per cent grade of iron concentrate has been drawn for comparison of grades of concentrate among the submembers. Figures for the silica content of these samples are not available, although other analyses indicate that, at this grade, the silica content of the concentrate approaches a maximum desired limit. Although the discussion of the various lithologic types of taconite will be in general terms of potential grade of iron concentrate, it is the concomitant silica increase with decreasing iron content that makes an iron grade much below 64 per cent marginal. It is proposed that the relic sedimentary bedding structures and the subsequent development of metamorphic grain fabrics normally control the amount of silica in the magnetic concentrate. It cannot be emphasized too strongly, however, that the detailed data presented here actually cannot be projected much beyond the immediate vicinity of either of the holes. In addition, it must be stressed that the results of Davis magnetic tube tests are relative, and only approximate the actual products obtained by mill practice.

These holes were selected to show what typical variations in the concentrating characteristics might be encountered within the high-grade metamorphic regions, shown by Figure 38 for hole 21, and within the intermediate-grade metamorphic regions, shown by Figure 39 for hole 32. Similar information from a low-grade metamorphic region, such as from hole 27E near Mesaba, is not currently available. Holes 21 and 32 were drilled by the Reserve Mining Company specifically for our stratigraphic

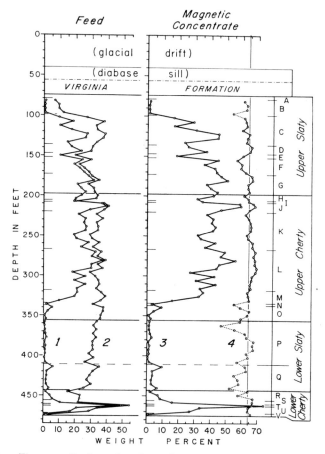

FIGURE 38. — The generalized stratigraphy and magnetic concentration characteristics of taconite core from hole 21. Curve 1 indicates the variation in weight per cent of iron occurring as magnetite in crude taconite feed from the numerous stratigraphic horizons. Similarly, the weight per cent of total iron in the crude taconite feed is shown by curve 2. The "recovery," or weight per cent of total iron recovered in the magnetic concentrate, and the "grade," or weight per cent of iron contained in that concentrate, are plotted on curves 3 and 4, respectively.

studies. The writers were also able to first mark all the stratigraphic submembers prior to having the cores assayed according to these units instead of arbitrary five-foot intervals. The actual test data are presented graphically in Figures 38 and 39, and certain values are summarized in Table 5 for the reader's convenience in visualizing the average "composition" of natural taconite feed.

In addition to its metamorphic history, the concentrating characteristics of a given taconite are dependent to a large extent on its original bedding structure or texture. Fortunately, most of the primary stratification structure is still obvious, and as explained in Chapter 1, this feature was employed to define a lithologic classification of five basic kinds

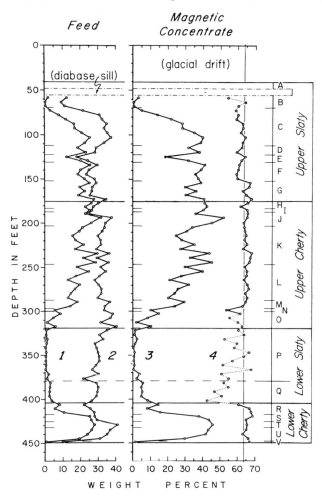

FIGURE 39. — The generalized stratigraphy and magnetic concentration characteristics of taconite core from hole 32. Numbered curves are as described in the legend of Figure 38.

TABLE 5. AVERAGE COMPOSITION OF NATURAL TACONITE FEED

Member	Thickness	% Fe (total)	% Fe (in mt.)
Upper Slaty*			
Hole 21	118.5	26.5	16.9
Hole 32	119.5	27.0	17.0
Upper Cherty			
Hole 21	158	34.1	21.5
Hole 32	144	30.9	16.9
Lower Slaty			
Hole 21	88	29.9	1.4
Hole 32	85	29.3	1.3
Lower Cherty			
Hole 21	31	24.6	11.9
Hole 32	45	30.8	19.2

*Not including the carbonate-rich submember A at the top of the formation.

of taconite: *massive, layered, laminated, shaly bedded,* and *shaly*. In general, these different kinds of taconite differ in amenability to magnetic concentration. Submember A of the Biwabik formation, generally a calcite marble, contains a trace of iron in its constituent lime-silicates, although it is obviously eliminated from consideration as a source of iron.

Massive varieties of taconite contain few magnetite-rich bedding structures, and the lean magnetite content consists largely of fine grains disseminated throughout a quartzose to silicate-rich matrix. In numerous instances the magnetite is so finely divided, particularly within "dusty granules" (Fig. 40, A) as in some parts of unit G, that it produces numerous middling grains that locally reduce the grade of the concentrate. In some places, however, the magnetite in the granules (Fig. 40, B) is sufficiently recrystallized so that the middling problem is largely eliminated. In many of these places, as in units E, I, and J, the massive taconites are generally too lean to consider as milling material. In general, few massive taconites contain significant amounts of recrystallized magnetite-rich granules, as occur locally in units J, T, U, and to a lesser extent in parts of unit G, but they warrant consideration for concentration. Units G and J are currently being mined in the Mitchell pit. In a few places, as in parts of units N and O, the magnetite-rich granules are commonly recrystallized, though significant quantities of magnetite have been consumed in the formation of fayalite, ferrohypersthene, and cummingtonite, and can no longer be recovered by simple means. Much of the residual magnetite is finely divided and poikilitically distributed within the silicates, the presence of which commonly causes a submarginal grade of concentrate.

Another variety of rock from the iron formation, shaly taconite, occurs largely within the Lower Slaty member. This rock is normally silicate-rich and is essentially a variety of silicate-rich massive taconite that has a thin-bedded or laminated aspect, as explained in Chapter 1. Submember P of the Lower Slaty member is composed largely of silicate-rich massive and shaly taconites in which nearly all of the previously existing magnetite has been consumed (compare curves 1 and 2 of Figs. 38 and 39) in the formation of fayalite, ferrohypersthene, and cummingtonite. Similarly, the dark, carbonaceous, thinly-bedded unit Q (the "Intermediate Slate") contains little residual magnetite. Because no magnetic iron minerals except traces of pyrrhotite remain in the Lower Slaty member, the only conceivable method of recovering its iron content would be by some expensive process of direct reduction. It is interesting to note, however, that the total iron content is of the same order of magnitude as in the remainder of the formation.

The other three varieties of taconite contain appreciable amounts of magnetite-bearing bedded taconite-strata (see Chapter 1 for detailed discussion). Paradoxically, the variety of rock called layered taconite, defined as consisting of from 10 to 40 per cent by volume of bedded taconite-strata, normally contains fewer and thinner magnetite-bearing bedded zones, but this variety generally yields the most satisfactory concentrate of the three. This is mainly because the initial state of aggregation of the magnetite collected during sedimentation was such that the layered beds, particularly upon simple recrystallization, became relatively pure and free of admixtures of associated quartz and later metamorphic silicates.

Typical grain fabrics found among the magnetite-rich layered taconites are shown in Figure 41 (A and B). Taconites with these grain fabrics yield acceptable concentrates upon magnetic concentration, and rocks of this kind constitute the great bulk of material currently being mined from units H, K, L, and M of the Upper Cherty member in the Mitchell pit. Similar rocks from units R, S, and U might be mined in the future, when it becomes economically feasible to attempt to mine the thin Lower Cherty member in the Eastern Mesabi district. In numerous places, however, thin zones containing locally abundant quantities of silicates occur within the magnetite-rich beds of the layered taconites. Where this occurs, the silicates are intimately intergrown with, or poikilitically enclose, the magnetite grains to form middling grains, and the quality of the magnetic concentrate can abruptly change to a marginal or undesirable grade.

To be sure, the magnetite-rich beds of the units listed directly above do not consist entirely of layered taconite alone, and minor amounts of laminated and shaly bedded taconites can also occur on a small scale to bring about a diminution of concentrate grade. Granule-bearing quartzose layers commonly are interbedded with the magnetite layers of all the bedded varieties of taconite. If these granules are not recrystallized, as explained above, they can contribute middling grains. Fortunately, in many of the occurrences of layered taconites the magnetite-quartz granules have been largely consumed in the formation of silicates which can be easily rejected during the cobbing stages of concentration. Occurrences of intergrown silicates in the magnetite-rich layers, admixtures of other varieties of taconites, or of abundant "dusty" magnetite granules

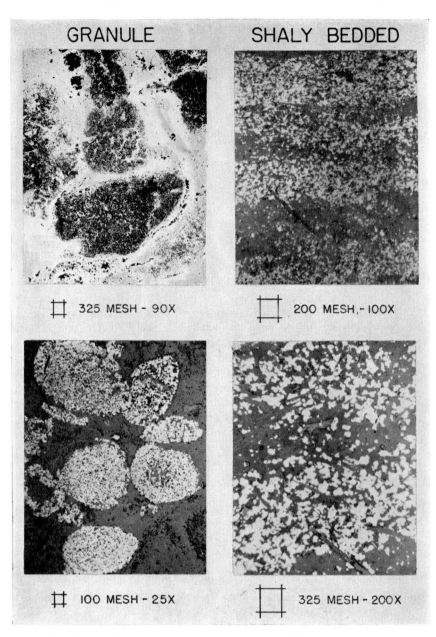

FIGURE 40. — Grain fabrics in magnetic taconite. A (upper left): Granule structures consisting of "dusty" magnetite in quartz. Typical fine-grain fabrics seen in thin section at 90X. The square shows the opening in a 325 mesh sieve at this magnification. B (lower left): Recrystallized granule structures rich in magnetite. Coarsened fabrics as seen in polished section at 25X. Black spots are ground-out voids. The square grid shows a 325 mesh sieve opening at this magnification. C (upper right): Shaly bedded taconite. Shaly bedded aggregates of fine magnetite anhedra as seen in polished section at 100X. The size of opening of a 100 mesh sieve at this magnification is indicated by the square. D (lower right): Shaly bedded taconite. Same view as above but at 200X. The square indicates the size of opening of a 325 mesh sieve at this magnification.

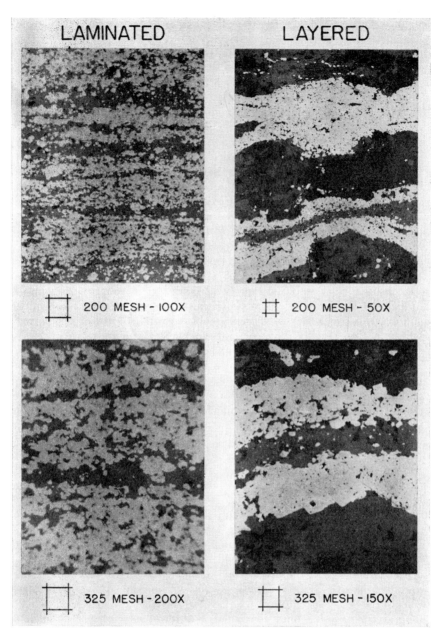

FIGURE 41. — Grain fabrics in magnetic taconite. A (upper left): Laminated taconite. Laminated aggregates of magnetic grains as seen in polished section at 100X. 200 mesh sieve opening at this magnification is shown by the square. B (lower left): Laminated taconite. Same view as above but at 200X, along with the size of opening of a 325 mesh sieve at this magnification. C (upper right): Layered taconite. Layered aggregates of coarsened magnetite grains as seen in polished section at 50X. 200 mesh sieve opening at this magnification shown by the square. D (lower right): Layered taconite. Same view as above but at 150X, along with the size of opening of a 325 mesh sieve at this magnification.

produce thin local zones in the taconite that yield marginal grades of concentrate, as indicated by the sharp decreases in grade (curve 4 of each graph, Figs. 38 and 39, corresponding to the units mentioned in the preceding paragraph). Unit B is also largely layered taconite; in this instance, however, the iron-rich layers are mainly diopside and none of the contained iron (units) can be effectively removed, at least by magnetic concentration.

In the group of iron-formation rocks classified as laminated taconites, the aggregates of magnetite grains that form the thin magnetic beds did not initially accumulate into particularly pure zones of magnetite. Instead, these magnetite aggregates are commonly intercalated with quartzose zones or layers of nearly equal thickness, as shown in Figure 41 (A and B). The grain-size distribution of the magnetite in this type of taconite is such that at a theoretical particle size of 325 mesh a significant number of grains can consist of interlocked middling grains. Consequently, a marginal to nearly satisfactory quality of concentrate is likely to result from rocks consisting largely of laminated taconite; for example, units C and D. The portions of these units that yield a suitable product contain laminated zones that are slightly recrystallized to a more favorable grain size. Locally, however, interlocking of magnetite with silicate grains formed during metamorphism can add to the middling-grain problem of maintaining grade, as is shown by the sharp drops in concentrate grade (curve 4 of Figs. 38 and 39) corresponding to a few thin zones in units C and D. Consequently, although the average magnetite-bearing bedded taconite-strata of the laminated taconites are generally thicker than those found in the layered taconites, the primary bedding fabric governing the original distribution of the magnetite aggregates greatly influenced the ultimate concentrating characteristics of two groups of taconites that are somewhat similar mineralogically and which commonly contain the same order of magnitude of total iron and of iron as magnetite.

Unit F, consisting largely of shaly bedded taconites, normally yields the poorest grade of magnetic concentrate obtained from any unit containing appreciable quantities of magnetite. Although the total iron and magnetic iron content of shaly bedded taconite ranges only about 5 to 10 per cent less than the layered and laminated varieties, the finely disseminated nature of the magnetite grains within the thinly bedded taconite-strata of the rock prevents effective recovery. Nearly everywhere, fine-grained cummingtonite is intimately intergrown with the magnetite, and in the few places where the magnetite-bearing beds are slightly recrystallized without forming appreciable numbers of silicates during metamorphism a substantial number of grains are still much smaller than 40 to 50 microns and cannot be easily liberated and collected. Figure 40, C, D, shows typical grain fabrics likely to be found among the relic bedding structures of the shaly bedded taconites. These thin beds, slightly recrystallized, contain only minor amounts of silicates; nevertheless, at a theoretical particle size of 325 mesh, crushed fragments of the rock would consist mainly of interlocked middling grains.

As often as not, a fundamental study concerning basic geological problems will yield extremely practical information. The academic approach to the stratigraphy and studies of metamorphic zoning in this district necessitated delineation of subtly different rock types and stratigraphic submembers. A clear correlation between lithologic types and the concentrating characteristics of these rocks was later noted, as discussed above. Many of the proposed submembers are thin and most consist, essentially by definition, largely of one or two basic taconite types. As a result of initial variations during sedimentation, zones of good and poor "concentratability" are laid one on top of the other. During large-scale mining, some zones of poor material must necessarily be taken in with the large bulk suitable for milling.

The structure of the nearly flat-lying rocks of the Biwabik formation is basically simple, uncomplicated by any extensive development of faults and folds. However, broad folds and flexures of small amplitude occur in the portion of the district now being mined. These small-scale structures present no difficulties in nonselective mining except near the top of the stratigraphic zones being mined. Here, a small down-warping flexure of a few tens of feet can drop unit F, of poor milling material, directly into the westward path of the mining. Such features make it necessary to divert the direction of mining northward, at least temporarily, to remain within suitable milling material.

Normally it is desired to have the lower limit of rock breakage and mining set at the basal strata of the Upper Cherty member, because the underlying strata of the Lower Slaty member are silicate-rich and lean in magnetite. Silicates, however, are also locally abundant in units M, N, and O (of the Upper Cherty), which have had most of their available magnetite consumed in many places; much of the remnant magnetite is unattainable at suitable grade because of its finely disseminated nature in poikilitic silicates. Consequently, during the early stages of mining, small local regions within the basal units of the Upper Cherty member were encountered which added little but middling grains to the concentrate or eventual bulk to the tailings product. Although these areas of rocks were unavoidably broken during large-scale mining, current detailed core drilling programs have essentially eliminated this problem by locating the basal mining limits of the pit prior to blasting.

SUMMARY

The Lower Cherty member of the iron formation is essentially a thin wedge of sediments, ranging from 30 to 50 feet thick, in the Eastern Mesabi district. The member is not being mined on the Reserve Mining Company property, but to the west — beyond the probable fault — it thickens to about 136 feet near Mesaba and is currently being mined by the Erie Mining Company. The basal unit of the member does not contain sufficient iron as magnetite to concentrate. Although the uppermost unit consists largely of layered taconite, the local intimate association of silicates with the magnetite has resulted in material that is difficult to bring up to grade. The middle three units, consisting largely of magnetite-rich granule-bearing massive and layered taconites, contain silicates that developed largely within the quartzose strata of the rocks, which did not lead to the development of finely interlocked magnetite-silicate middling grains in the magnetic portions of the rock. It is therefore possible that these units of the Lower Cherty member on the Reserve property might be amenable to concentration, but the member cannot be developed commercially at present because only about 15 to 25 stratigraphic feet could be considered successful sources of milling material.

The Lower Slaty member, consisting of massive silicate taconite and shaly silicate taconite, contains only slightly less total iron than the adjacent Cherty members. Most of this iron, however, is combined in the abundant ferrous silicates that are found throughout the district. The carbonaceous unit Q is essentially devoid of magnetite, although minor amounts of magnetite are locally present in unit P. In this latter unit some of the residual magnetite is slightly coarsened, but much of the magnetite is merely poikilitically disseminated as fine grains in the silicates, thus forming numerous middling grains. It is not likely that the Lower Slaty member, which averages about 86 feet thick in the district, will be used as a source of iron in the immediate future.

The Upper Cherty member, averaging about 140 feet thick in the Eastern Mesabi district, includes most of that part of the Biwabik formation currently being mined in the Mitchell pit of the Reserve Mining Company. It is also being mined to a lesser extent by the Erie Mining Company in a pit just northeast of Mesaba, where the member is about 246 feet thick. The middle units of the member, J, K, and L, contain the bulk of the material mined in the Mitchell pit. Although there are minor zones containing intimately associated magnetite and silicates, the magnetite aggregates have collected mainly into relatively rich layers, lamellae, granules, and pebbles within generally quartzose material. Earlier "dusty" magnetite-quartz granules of the quartzose layers have been largely reconstituted into silicates during metamorphism, thus largely eliminating a serious middling-grain problem. The lower three units, M, N, and O, similar to those just described, contain relatively subordinate amounts of magnetite, of which a large portion was consumed in the formation of the locally abundant silicates, as occurred in the Lower Slaty member. The algal unit I is typically lean. H, the uppermost unit of the member, contains rich layers of magnetite, but they are characteristically associated with the local development of hornblende and actinolite. Again the association of magnetite and a silicate that has practically replaced it produced magnetic concentrates that are generally marginal.

The Upper Slaty member, averaging about 125 feet thick, is normally more distinctly layered and thinly bedded than other members of the iron formation, due to the predominance of laminated and shaly bedded taconites. The lowest unit, G, of the member consists largely of recrystallized, magnetite-rich granules and lamellae in the quartzose matrix, and — because of fair concentrating characteristics — is currently being mined along with units of the Upper Cherty in the Mitchell pit. Another thin zone of massive taconite, unit E, or the "septaria zone," is avoided in mining because it adds little but middling grains to the concentrate. Unit F, consisting largely of shaly bedded taconite is not being mined at present because of the difficulties encountered in liberating and separating the magnetite grains from the intimately associated silicates. This unit might be utilized in the future, however, as the technology of fine grinding and magnetic recovery of extremely fine particles is perfected. Units C and D, containing mainly laminated taconites, might be capable of producing marginal concentrates with present milling procedures, although these rocks are unexposed in the area currently stripped for mining and are relatively far removed stratigraphically (and thus geographically when one considers the shallow dip of these rocks) from the units now being mined. It is likely that these units could also be exploited when grinding and recovery techniques are perfected. A and B, the two uppermost units of the formation, are essentially devoid of magnetite, and their minor iron content is tied up in lime-silicates, thus making them unlikely sources of iron.

REFERENCES

Broderick, T. M., 1919, Detailed stratigraphy of the Biwabik iron-bearing formation, East Mesabi district, Minnesota: Econ. Geol., 14:441–451.
Chester, A. H., 1884, The iron region of northern Minnesota: Minn. Geol. and Nat. Hist. Survey, Ann. Rept., 11:155–167.
Eames, H. H., 1866, Report of the state geologist on the metalliferous region bordering on Lake Superior: St. Paul, Minn., 21pp.
Flaschen, S. S., and E. F. Osborn, 1957, Studies of the system iron oxide-silica-water at low oxygen partial pressures: Econ. Geol., 52:923–943.
Grout, F. F., and T. M. Broderick, 1919, The magnetite deposits of the Eastern Mesabi range, Minnesota: Minn. Geol. Survey Bull. 17.
———, et al., 1951, Precambrian stratigraphy of Minnesota: Geol. Soc. Am. Bull., 62: 1017–78.
Gruner, J. W., 1924, Contributions to the geology of the Mesabi range: Minn. Geol. Survey Bull. 19, 71pp.
———, 1946, Mineralogy and geology of the Mesabi Range: Office of the Commissioner of Iron Range Resources and Rehabilitation, St. Paul, Minn., 127pp.
Gundersen, J. N., 1958, The stratigraphy and mineralogy of the metamorphosed Biwabik iron-formation, Eastern Mesabi district: Ph.D. Thesis, University of Minnesota, 180pp.
———, and G. M. Schwartz, 1959, Metasomatic veins in the Biwabik iron-formation, Minnesota: Geol. Soc. Am. Bull. 70:1613.
———, 1960a, Lithologic classification of taconite from the type locality: Econ. Geol., 55:563–573.
———, 1960b, Stratigraphy of the Eastern Mesabi district, Minnesota: Econ. Geol., 55:1004–1029.
———, and G. M. Schwartz, 1961, Magnetic taconites of the Eastern Mesabi district, Minnesota: Mining Trans. Am. Inst. Min. Eng., 222: 227–233.
James, H. L., 1954, Sedimentary facies of iron-formation: Econ. Geol., 49:253–293.
Leith, C. K., 1903, The Mesabi iron-bearing district of Minnesota: U.S. Geol. Survey Mon. 43, 316pp.
Richarz, S., 1927a, Grunerite rocks of the Lake Superior region and their origin: Jour. Geol., 35:690–707.
———, 1927b, The amphibole grunerite of the Lake Superior region: Amer. Jour. Sci., 14:150–154.
Smith, J. R., 1957, Reconnaissance in the system $FeO-Fe_2O_3-SiO_2-H_2O$: Carnegie Institution of Washington, Annual Report, 1956–1957, pp. 230–231.
Winchell, N. H., 1879, Sketch of the work of the season 1878: Minn. Geol. and Nat. Hist. Survey, Ann. Rept., 7:9–25.
———, 1881, Preliminary list of rocks from the crystalline formations of northern Minnesota: Minn. Geol. and Nat. Hist. Survey, Ann. Rept., 9: 100–204.
———, 1889, The crystalline rocks of Minnesota: Minn. Geol. and Nat. Hist. Survey, Ann. Rept., 17:5–74.
———, and H. V. Winchell, 1891, The iron ores of Minnesota, their geology, discovery, development, qualities and origin, and comparison with those of other districts: Minn. Geol. and Nat. Hist. Survey, Bull. 6, 430pp.
Winchell, A. N., and H. Winchell, 1951, Elements of Optical Mineralogy: Part II, John Wiley and Sons, Inc., 551pp.
White, D. A., 1954, The stratigraphy and structure of the Mesabi Range, Minnesota: Minn. Geol. Sur. Bull. 38, 92pp.
Wolff, J. F., 1917, Recent geological developments on the Mesabi Iron Range: Trans. Am. Inst. Min. Eng., 56:142–169.
Yoder, H. S., 1957, Isograd problems in metamorphosed iron-rich sediments: Carnegie Institute of Washington, Annual Report, 1956–1957, pp. 232–237.

INDEX

Actinolite: described, 82–83
Almandine (almandite), 29, 30, 91
Albite: 89–90; in pegmatite, 71
Algal structure, 28, 51, 52
Alkali feldspar: described, 89–90
Amphibole group: retrograde, 74; described, 82–90
Andesine: described, 90–91
Andradite, 29, 68, 91
Ankerite: 48, 60, 103; in veinlets, 64, 66, 72
Anthophyllite, 118, 122
Apatite, 66, 72, 73, 74, 82, 86, 102, 122

Babbitt, 5
Barite, 28, 99
Barium, 73
Biotite: in hornfels, 68; in gabbro, 72; described, 91–92
Birch Lake, 3, 5, 6
Biwabik iron-formation: length, 6; composition, 8; division into submembers, 14, 22; columnar section, 15; contact with Pokegama quartzite, 19–22; general features, 22–24; red basal member, 25; primary silicates, 118; structure, 133
Biwabik iron-formation, submembers: letter designations of, 24; A, 67–68; B, 66–67; C, 64–66; D, 62–64; E, 60–62; F, 58–60; G, 56–58; H, 53–55; I, 51–53; J, 48–51; K, 45–48; L, 42–45; M, 39–42; N, 38–39; O, 36–38; P, 34–35, 105; Q, 31; R, 28–31; S, 27–28; T, 27; U, 25–27; V, 25; concentrating characteristics, 128–135

Calcite: 74; analysis, 76; occurrence, 79, 81, 102
Calcium: in metasomatism, 73; introduced, 75
Carbon, 77
Carbonates: 81; fluids introducing, 79
Chalcopyrite: 29, 38, 41, 62, 67, 68, 71, 73, 74, 86; occurrence described, 77
Chert: usage defined, 32
Chert conglomerate: relic, 64
Chlorite: described, 95
Chrysotile, 72
Climate, 5

Clinohypersthene: described, 92–93
Conglomerate zones: in submember Q, 31
Cordierite, 69
Covellite, 77, 116
Cummingtonite: 83–86; defined, 74; analysis, 76; mode of formation, 123

Davis, E. W., 4, 124
Davis tube: used, 125. *See also* Magnetic tube tests
Diabase: dikes and sills described, 69–71; metamorphosed, 70; metasomatized, 70
Dikes, 69–71
Diopside: described, 93
Dolomite: near Mesaba, 121
Duluth gabbro: 1, 6; described, 71–72; thermal metamorphosis by, 114, 124

Eames, H. H., 3
Eastern Mesabi district: defined, 1, 6
Elements: introduced in iron formation, 77, 116
Ely greenstone, 6
Epidote, 67, 95
Erie Mining Company, 4, 36, 133, 134
Eulysite, 34, 35, 37

Fayalite: analysis, 76; described, 95–96; origin, 115; restricted occurrence, 115
Feldspar group, 89–90
Ferrohypersthene: analysis, 76; described, 94; adjacent to pegmatite veins, 116
Franz isodynamic separator: used, 76

Gabbro: 71, 72; metamorphosis by, 69. *See also* Duluth gabbro
Garnet group, 91
Geochronology, 12
Giants Range granite: 5, 6; described, 16–18; eroded surface, 17; outcrop at Old Babbitt, 17
Goldich, S. S., 76
Grain size: usage defined, 23
Granules: described, 23, 36, 56, 60; texture, 23; origin, 24; destroyed, 47; recrystallized, 55; relic granule structures, 79, 121

Graphite: 32, 33, 67, 77; with magnetite, 80; in intermediate slate, 80
Graphitic material: derivation, 115; in reducing environment, 115
Greenalite, 7, 118, 119
Gruner, J. W., 80
Gunflint district, 6

Hedenbergite: analysis, 76; described, 93–94; adjacent to pegmatite veins, 116
Hematite: 23; described, 78–79
Hisingerite: 28, 33, 35, 37, 38, 41, 44; in veinlets, 65; analysis, 76; described, 96–97
Hornblende: analysis, 76; described, 86–87
Hornblende+cummingtonite: described, 87–89
Hornfels: in Virginia formation, 68–69

Idocrase, 68, 97
Intermediate slate: 29, 31; concentrating characteristics, 129
Iron: concentrate grade, 125; content and recovery curves, 126, 127
Iron-oxide-silica-water system: 109–112; interpreted, 112–118
Isograds: 103–104; zones of, 103, 121

Jackling, D. C., 4

Knife Lake slate, 6
Kupfferite-grunerite series, 84

Laminated taconite, 132
Lamprophyres, 73
Layered taconite, 129
Lithostratigraphic units, 14, 16
Loellingite: 49, 73, 74, 102, 122; occurrence, 78
Lower Cherty member: described, 24–31; mining in, 133–134
Lower Slaty member: described, 31–35; concentrating characteristics, 128–129; mining possibilities in, 134

Mafic veins, 73, 117
Magnesium: introduced, 75; in cummingtonite, 84–85
Magnetic taconite: 128; in grain fabrics, 129–132
Magnetic tube tests: 125; curves derived from, 126–127
Magnetite: described, 79–81
Magnetite-rich: usage defined, 23
Marble, 67–69
Martite, 60
Massive taconite: 8, 32; concentrating characteristics, 128

Mesabi district: history of, 3–4
Mesabi Iron Company, 4
Mesabi Syndicate, 4
Metamorphic mineral assemblages: 102–104; in paragenetic sequence, 104–109
Metamorphic silicates, 102–103
Metamorphism: general process, 102; temperature, 114; thermal phase, 122; progressive, 122; metasomatic stage, 122–123; effects of, 122–123
Metasomatic activity: 73; minerals formed by, 122
Mica group, 91–92
Minerals: estimating amounts of, 23; methods of identifying, 76–77; reactions of, 112–113
Mines Experiment Station, University of Minnesota, 124
Mining: 124–135 *passim*; impeded by folds and flexures, 133; lower limits of, in Biwabik formation, 133. See also Mesabi district
Minnesotaite: 28, 29, 42, 51, 60, 63, 79, 103, 112, 118, 119; described, 97
Molybdenite, 73, 79, 116
Mountain Iron: ore discovery at, 4
Muscovite: described, 92

Nontronite: 34, 35, 56; in veinlets, 68; in gabbro, 72; described, 98

Oligoclase: 38; described, 90–91
Oxides, 78–81
Osborn, E. F., 114

"Paint Rock," 31
Palacas, James, 115
Paragenetic sequence: 104–109; phases of, 114; metasomatic stage, 116
Pegmatitic veins: 46; described, 72–75; elements introduced, 116; metasomatic metamorphism, 124
Peter Mitchell Pit, 3, 4, 36, 78, 81, 84, 87, 91, 94, 112, 128, 134, 135
Phosphate, 92
Plagioclase: 41, 43, 56, 57, 62, 99; in iron formation, 35; described, 90–91
Pleochroic haloes, 29, 37, 41, 43, 67, 72, 74, 88, 107
Poikilitic textures, 106
Pokegama formation: described, 18–22; contact with Biwabik formation, 19–22; in hole 27E, 21
Praseolite, 69
Precipitation, 5
Pyrite, 31, 72, 73, 78, 86, 116

Pyroxene group: described, 92–95
Pyrrhotite, 31, 32, 35, 38, 65, 67–69, 71, 74, 78, 116

Quartz: described, 98–99
Quartz-stilpnomelane-ankerite veinlets, 64

Reserve Mining Company: organized, 4; operations, 11, 124; holes drilled by, 126
Retrograde sequence, 103, 122
Riebeckite: described, 40, 89
Rock Analysis Laboratory, University of Minnesota, 76

Septaria structure, 60–61
"Septaria zone," 135
Shaly taconite: concentrating characteristics, 128–129, 132
Siderite, 29, 118
Silica: control of, in concentrate, 125
Silica-rich: usage defined, 23
Silicates: listed, 7; order of abundance, 82; described, 82–101; metamorphic, 102–103; decomposed, 119
Sills, 69–71
Soda-hornblende, 89
Sphene, 99
Stilpnomelane: 26, 45, 58, 60, 63, 116; in veinlets, 64, 67; retrograde, 73; described, 99–100
Styolites, 51
Sulfates: described, 81

Sulfides: described, 77–78

Taconite: defined, 6; type locality, 6; classification, 7–10; structural features, 8–10; concentrating characteristics, 126–127; average composition, 128; grain fabrics, 130
Temperature, 99
Thermal metamorphism, by gabbro, 72, 114
Titanite, 99
Tourmaline, 100
Troctolite, 71

Upper Cherty member: described, 35–55; magnetite content, 55, 64; in Mitchell Pit, 134
Upper Slaty member: described, 55; concentrating characteristics, 135

Vesuvianite, 97
Virginia formation: described, 68–69; effect of gabbro in, 69

Water: introduced in taconite, 75, 123; and dispersal of hedenbergite, 117
Winchell, H. V., 4
Winchell, N. H., 3
Wollastonite: 68; described, 100–101

Yoder, H. S., 113, 114, 119

Zircon: 29, 37, 41, 43, 73, 74, 88, 100; usage defined, 23